THE CFC STORY

THE CFC STORY

How America's Rural Electric Cooperatives Introduced Wall Street to Main Street

PATRICIA LLOYD WILLIAMS

**NATIONAL
RURAL
UTILITIES
COOPERATIVE
FINANCE
CORPORATION**

Woodland Park
2201 Cooperative Way
Herndon, Virginia
22071-3025

Printed in the United States of America

Design by Wickham & Associates, Inc.

Editorial management and production oversight by Off-Press Publishing.

ISBN 0-9645302-1-X
ISBN 0-9645302-0-1 (pbk.)

CFC's Mission

CFC is a not-for-profit cooperative whose mission is to provide its member utility systems—through their unified, collective strength—with an assured source of low-cost private capital and state-of-the-art financial services.

CFC's Role

- *To provide rural electric distribution systems with an assured source of capital at the lowest possible cost at whatever level is required*
- *To provide rural electric power supply systems an assured source of capital and, where possible, add economic value to market entry by power supply systems*
- *To provide resources, as required, to meet the evolving needs of the rural electric program for capital and financial services*
- *To operate and maintain CFC in a manner that will foster and develop cooperative principles of organization and philosophy*
- *To foster and support the maintenance and strengthening of the common goals and unity of purpose of the rural electric program*
- *To foster, develop and protect the cooperative form of organization in the utility sector*

C O N T E N T S

FOREWORD

When the members of the Long-Range Study Committee and the rural electric program created the National Rural Utilities Cooperative Finance Corporation (CFC), they created something very special. CFC is a unique financing organization, cooperative in form and nonprofit in nature. It was also the first private organization to offer mortgage-backed securities without a government guarantee. It enabled its members to do together things that they could not do individually.

What impresses me the most is that, after more than 25 years, the basic concept of CFC is still intact. The organization that was created in 1969 has adapted to meet the changing needs of the rural electric program and yet remained basically the same.

During the years since its creation, CFC has introduced the rural electric program to many new concepts, such as short-term financing and tax-exempt financing, and has provided tools, such as the commercial paper program, to enhance the member systems' financial management activities. Through good and bad economic times, it has made sure its members had access to the funds they needed on a competitive cost basis.

CFC created the National Cooperative Services Corporation and thus enabled its members to save millions of dollars through the use of innovative financing techniques such as leveraged leases. It expanded its efforts to support rural utility services by creating the Rural Telephone Finance Cooperative to provide loans and financial services to rural telephone systems. It supported the organization of the National Rural Telecommunications Cooperative to enhance telecommunications services in rural areas. It encouraged rural economic development activities through an associate member program.

Throughout this time, CFC has had one purpose and one purpose only: to provide its member systems, through their unified, collective strength, with an assured source of low-cost private capital and state-of-the-art financial services.

As the rural electric program prepares to enter the 21st century, this is still CFC's mission. It has been and always will be your organization. We hope you are proud of it.

Charles B. Gill
Governor and Chief Executive Officer
1979–1995

ACKNOWLEDGMENTS

The contributions to the formation of the National Rural Utilities Cooperative Finance Corporation (CFC) of such rural electric giants as J. K. Smith, Jerome S. Katzin and David A. Hamil are well-known. The contributions of Clyde Denton may not be as well recognized. As manager of information services for the Kentucky Association of Rural Electric Cooperatives, he assisted Smith in the work of the Long-Range Study Committee. He attended every meeting and wrote Smith's speeches and reports. At Smith's insistence, he reluctantly left Kentucky in January 1971 to join the CFC staff as its first information and member services officer. He retired from CFC in 1989. Much of his time during his later years with CFC was devoted to assembling information about CFC's early history. He conducted interviews with the individuals who played pivotal roles in the work of the Long-Range Study Committee and during the formative years of CFC, and he drew on his own memories as well to put together an account of how CFC came to be. Clyde's work made an enormous contribution to this book and is much appreciated.

Gary A. Gordy, who held several positions with CFC before his retirement in 1993, picked up where Denton left off and spent the last year of his career advancing *The CFC Story.*

On behalf of CFC, I would like to acknowledge the valuable contribution of the people who so generously participated in interviews with Clyde Denton or me (or, in some cases, both of us), including Truman Brandt, Milton A. Chase, Norman L. Clapp, Robert R. Edmiston, Paul Fox, David A. Hamil, Virgil Herriott, Jerome S. Katzin, Owen Manning, Roger Oresman, Robert D. Partridge, Louis A. Roland, Sam Shiozawa, Ira Shesser, Vincent P. Slatt, J. K. Smith and Nicholas J. Walsh as well as current members of the CFC staff, including Governor

Charles B. Gill, Richard B. Bulman, Martin R. Crowson, David J. Hedberg, Steven L. Lilly, John Jay List and Sheldon C. Petersen.

Thanks also to those who reviewed the manuscript in its draft form and provided useful criticisms and to my friends at CFC who interrupted their work to do research for me.

I especially note the help of Ira Shesser, who even in retirement retains most of the corporate memory of CFC. He was always available to answer questions and suggest new avenues of research, and he labored beyond the call of duty by critiquing the draft not once but twice. Thanks, too, to Milt Chase, who knows as much about generation and transmission cooperatives as Ira does about CFC.

CFC Director of Communications Michael J. O'Brien was invaluable in serving as a liaison with the CFC staff and in coordinating the production process. He provided editorial guidance when I needed it and never stopped trying to teach me the difference between "which" and "that."

The contribution of my husband Andrew W. Williams should be noted. He asked me how many pages I had written every single day and urged me on to greater production when he didn't think it was enough.

I thank Governor Gill for the wonderful opportunity to work on this project. Most of all, I thank the members of the rural electric program for creating a very special organization and allowing me to be a part of its history.

Patricia Lloyd Williams

PART 1 WHY CFC

By the late 1950s, the rural electric program's need for funds exceeded what Congress was willing to provide through the traditional loan program administered by the United States Department of Agriculture's Rural Electrification Administration. At the same time, changes in rural electric cooperative service territories resulted in a rate of growth in demand for electricity higher than that of investor-owned and municipal utilities. The leaders of the rural electric program recognized the need to develop a supplemental financing program to ensure that adequate funds were available to meet the program's growing needs. The National Rural Electric Cooperative Association (NRECA) first attempted to establish a rural electric bank that would initially be supported by the government and ultimately become an independent agency. When the legislation needed to create the bank failed because of massive opposition from investor-owned utilities and resistance from a minority of cooperative systems, NRECA established the Long-Range Study Committee to develop recommendations on alternative approaches to meeting the long-range financial needs of the rural electric program. On March 17, 1969, representatives to the NRECA annual meeting voted to establish a cooperative self-help financing organization. That organization was the National Rural Utilities Cooperative Finance Corporation—CFC.

Highlights

- *In 1958, Rural Electrification Administration (REA) Administrator David A. Hamil and the Kentucky Rural Electric Cooperative Corporation began to call attention to the need to consider additional sources of financing for the rural electric program.*

- *In 1963, Buckeye Power, Inc., became the first cooperative system to access non-REA sources of funds for its participation with Ohio Power Company in the ownership and operation of the Cardinal Station, two 615-megawatt coal-fired units.*

- *At its 1964 annual meeting, the National Rural Electric Cooperative Association (NRECA) membership endorsed a proposal to study the feasibility of obtaining funds in the private capital markets.*

- *The Rural Electric Financing Report, issued in July 1965, recommended the creation of a Federal Bank for Rural Electric Systems. Legislation to establish the bank failed.*

- *In August 1967, NRECA established the Long-Range Study Committee, chaired by Kentucky statewide association manager J. K. Smith, to devise a plan for meeting future financing needs.*

- *At NRECA's March 1969 annual meeting, the Long-Range Study Committee recommended and the membership approved the establishment of a nonprofit, self-help cooperative financing organization, CFC.*

Sometimes an important event goes unnoticed when it happens. Only later does anyone recognize that something historic has occurred.

That was not the case with the incorporation of the National Rural Utilities Cooperative Finance Corporation (CFC) on April 10, 1969. Those present were taking part in the most significant event in rural electrification since May 11, 1935, when President Franklin D. Roosevelt established the Rural Electrification Administration (REA), and they knew it. They had worked hard for the moment.

"The members of the board were very conscious of the fact that they were opening the doors to a new era," said Ira Shesser. Shesser was involved in the work that led to the creation of CFC and served as its first loan officer. "They were all leaders who were very aware of the problems facing the program. They had been working on this problem for nearly two years. They understood the purposes and objective of the organization; they were sufficiently familiar with the rural electric program that they knew what had to be done; they knew how the organization had to be structured; and they knew what steps had to be taken to turn it from a paper organization into an operating entity."

There was no particular celebration. The incorporators signed the documents that needed to be signed and went back to the work of organizing the first nonprofit cooperative self-help financing institution in rural electric history.

The incorporation of CFC was an event that changed virtually every aspect of the rural electrification program, but the CFC story does not begin on April 10, 1969, any more than it ends there.

When President Roosevelt established REA, it was part of his overall program for unemployment relief, but it soon became apparent that a more extensive effort was necessary. The Rural Electrification Act of 1936 gave REA the statutory authority to finance the construction and operation of electric generation, transmission and distribution facilities to provide power to consumers in rural areas who did not have access to electric service. In 1949, REA's mandate was expanded to include authorization to loan funds for telephone service in rural areas. Congress approved annual appropriations to fund REA's loan programs.

During the post–World War II years, demand for loan funds routinely exceeded budget requests and authorizations. Then came stagnation during the Eisenhower administration, which was marked by Congressional battles over appropriations, interest rates and sources of funds. Legislation proposed in 1953 to increase the interest rate to 4 percent, the first effort to change the rate since it was fixed at 2 percent in 1944, was defeated, but it raised concerns. The succeeding Congresses saw attempts to increase interest rates to reflect the government's cost of funds and to place restrictions on the use of funds for generation and transmission facilities.

Electricity Spurs Growth

At the same time, the cooperatives' need for capital was increasing as rural areas became more suburban, in part because of the availability of electricity provided by rural electric systems. "Rural electric lines made the rural areas very attractive places to live. Housing followed the service," said J. K. Smith. Smith was general manager of the Kentucky Rural Electric Cooperative Corporation, a statewide association, and later became the first governor and chief executive officer of CFC.

So-called "string towns" developed along the lines. As soon as co-op lines were built to city limits, developers started building houses along them. "This was all due to the fact that electricity and other utility services were being made available. This was taking place throughout the nation. It was apparent that more and more capital would be required by the rural electric systems," Smith said.

By 1958, a few program leaders were beginning to speak out about the need to look at other sources of financing. REA Administrator David A. Hamil addressed the issue at industry meetings. "It became apparent to me that if REA borrowers were going to do what they told me they wanted to do—and what I hoped they would do—they were going to need more money than we were ever going to be able to convince Congress to make available to us at the 2 percent interest rate. There was a lot of opposition to the idea when I made those comments," he said.

It was also in 1958 that the Kentucky statewide adopted a resolution at its annual meeting calling on the National Rural

Electric Cooperative Association (NRECA) to consider establishing a study committee to review the program's long-term financing. "Many in the program felt that although REA was providing long-term financing at 2 percent, the time would come when the federal government would not provide all the financing we would need," Smith said. "We felt that we shouldn't expect the federal government to take care of our financing needs indefinitely. Rather, it was our opinion that we should look outside the federal government for a way to help ourselves in accordance with our ability to do so. We considered this as an act of good citizenship." The Kentucky group wanted to be sure that there would be a way to finance the program in the future if the government could not provide adequate capital.

Neither Hamil's exhortations nor the Kentucky group's resolution received much support. "There were times when it was questionable whether I was welcome at some of the meetings I attended," Hamil said.

The Kentucky resolution died. "It received very little support, because 2 percent money was a sacred issue. The general feeling was that we should leave things as they were until we were forced to do something," Smith said. "We in Kentucky became pretty unpopular throughout the rural electric program."

Hamil quietly continued his efforts. He asked both REA legislative consultant Charles U. Samenow and Fred Mynatt, assistant general counsel for the Department of Agriculture, to develop outside financing plans. Both plans proposed operations that were similar to the Farm Credit System (FCS), which used government-provided equity to raise funding in the private markets. Hamil wanted to bring additional capital into the program and develop an independent financing system, but he left office with the change of administration before he could fully develop these ideas.

Clapp Emphasizes Generation

Norman M. Clapp was appointed REA administrator by President John F. Kennedy in 1961. He saw much that needed to be done. "The Eisenhower administration has been characterized as a caretaker administration on many fronts. I think the truth of this characterization showed up in the REA program," Clapp said.

Construction by distribution systems increased rapidly after World War II, but issues affecting the long-term competitiveness of the systems, such as territorial protection and power supply, received less attention. Clapp was particularly concerned with power supply. Some distribution systems had banded together to form generation and transmission cooperatives (G&Ts) to supply their power needs. The G&Ts were based on the concept that the distribution systems would benefit from building and sharing one large plant rather than individually constructing several small ones. The G&T member systems entered into all-requirements wholesale power contracts, agreeing to purchase all of their power needs from the G&T and also agreeing to pay all of its costs.

Nationwide, most of the rural electric program's power came from another source. In 1962, for example, 19 G&T systems and 58 distribution systems generated a portion of their power needs. The total energy generated accounted for only 17 percent of the electric energy input of all REA borrowers for the year. The remainder was purchased from federal sources, state or municipal agencies, or investor-owned utilities. The private power companies were generally willing to sell power to the co-ops and even to offer a competitive rate, but the service terms and conditions often contained onerous restrictions on the loads cooperatives could serve. Many co-ops had a strong desire to escape the controls of the investor-owned utilities and develop their own resources.

REA policy restricted generation and transmission loans to cases in which no power was available or in which a G&T project could provide power at a lower cost. REA also required that projects show savings and that the full capacity be utilized by Rural Electrification Act beneficiaries within 10 years. These provisions had the effect of limiting the size of plants the G&Ts could build, which prevented them from enjoying the benefits of the economies of scale larger plants offered. Another objection to this policy was that it gave investor-owned utilities the opportunity to block projects by temporarily offering a lower rate, and many did.

"In REA, there was a concern that if you didn't have some restrictions, all the money would be going to G&Ts," said Milton A. Chase, CFC engineering officer from 1972 to 1980. Chase joined REA in 1937, specializing in bulk power supply issues. He helped to develop REA's policy on G&T loans. "We felt that if a co-op were able to get a lower cost of power under a satisfactory wholesale power contract, it would be wise not to divert funds for a G&T."

By the late 1960s, there was an apparent limit on funds available for REA electrification loans.

Clapp petitioned the Kennedy administration for support of an expanded G&T program. President Kennedy could continue the REA program as it had been for the last decade, which would not be controversial, he said, or it could address the power supply problems and secure generation and transmission for the cooperatives. "Those needs are very controversial, but they will distinguish this administration if they are done and will be among the efforts for which this administration will be remembered," he told Kennedy's advisors.

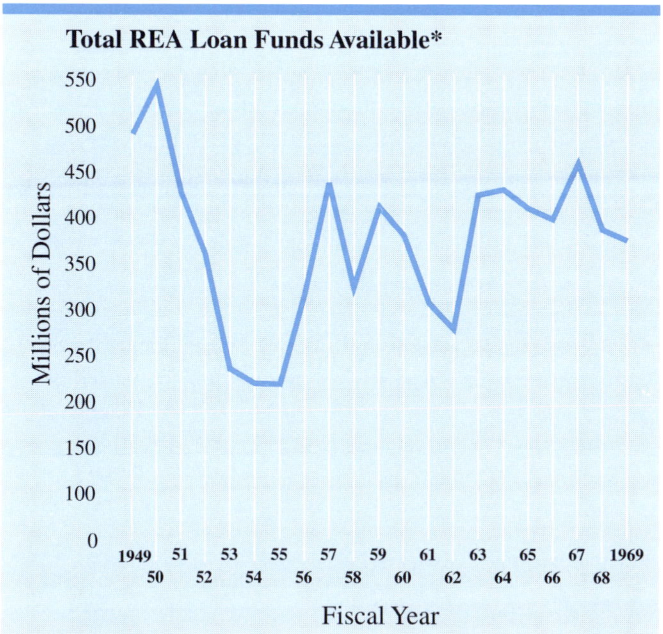

*Includes regular and reserve loan authorizations, carryover from prior year and rescission of prior loans.

The second alternative had a special appeal to the urban Kennedy, who had been criticized at times by rural leaders for his lack of knowledge of rural issues. "The word came back that the President had said, 'Go to it!' so I went to it," Clapp said. He revised REA's G&T loan policy to permit loans in cases where the terms and conditions under which a system could purchase power restricted the loads the co-op could serve or were otherwise unacceptable.

Clapp faced two immediate challenges in implementing his ideas. One was the apparent ceiling on REA appropriations.

The other challenge was opposition to the 2 percent interest rate. Opponents of the program used the interest rate issue to their advantage. "They said they were not against the REA program or the funding of it, but they contended that there was no justification for giving the REA cooperatives 2 percent money when people elsewhere and for other purposes were paying more," Clapp said.

Parity for Rural Consumers

Clapp justified the need for lower cost funds on the basis of parity of rates and services. He said rural consumers should have the same availability, quality and cost of electric service as urban consumers. "The job of rural electrification has special handicaps," he said. One was low density. In 1961, for example, REA borrowers served only 3.3 consumers per mile of line compared to more than 30 for investor-owned utilities. Thus, co-ops had to build many more miles of line than an urban utility to serve fewer consumers. Also, in part because of the restrictive terms and conditions in power purchase agreements with investor-owned utilities, cooperatives served few attractive industrial and commercial customers. Cooperatives did not earn nearly as much revenue on a unit of investment as urban utilities. "You had to come up with some equalizers if you were going to be able to deliver electric service to rural people on a par with that which city people enjoyed and at a cost comparable to what city people were required to pay."

There was also opposition to Clapp's parity concept from within the program. "Some of the real zealots, particularly out in the Plains states, believed that rural electrification should be a yardstick for the nation's utility industry. They weren't shooting for parity. They wanted to beat everybody else," he said.

Buckeye and Ohio Power Work Together

While Hamil and Clapp were thinking about ways to get more funding for the rural electric program, rural electric systems in Ohio were gaining access to the private capital markets and taking control of power supply resources in a unique and precedent-setting way. Buckeye Power, Inc., an Ohio G&T, became the first cooperative to finance generating capacity through private sources rather than through REA. The project was the Cardinal Station, two 615-megawatt coal-fired units, constructed jointly by Buckeye and Ohio Power Company. In addition to the generating capacity, the agreements between Buckeye and Ohio Power gave the cooperatives access to transmission facilities and wheeling. It demonstrated that a cooperative could successfully gain access to the private capital markets. And it introduced Jerome S. Katzin of Kuhn, Loeb & Company—the genius behind the

CFC concept—to the rural electric program.

Buckeye was established in 1959 to provide power to 30 member distribution systems in Ohio. The Buckeye members provided service to more than 135,000 consumer-members in 74 of Ohio's 88 counties. At the time Buckeye was formed, the members purchased wholesale power from seven investor-owned utilities. Their needs for both generation and transmission were significant.

Construction on Buckeye Power, Inc.'s, Cardinal Station was the first in the rural electric program to be financed using funds provided by the private capital markets.

Some of the Buckeye members purchased power from Ohio Power, a subsidiary of American Electric Power Company, Inc. (AEP), a public utility holding company that included six interconnected utility affiliates. Ohio Power, AEP's largest affiliate, served 485,000 customers in 53 Ohio counties.

AEP President Philip Sporn was unusual among investor-owned utility executives of the time in that he saw the cooperatives as an opportunity, not a threat. Sporn approached Buckeye's leaders with a project idea on February 16, 1962.

"We believe rural electric cooperatives are here to stay and have grown to such stature that they should render full utility service," he told them. "Perhaps we could build generation together, using some of the other investor-owned companies' existing transmission for wheeling. This could achieve the rural electric objective of providing to members an adequate supply of low-cost dependable electric energy far into the future." It would also relieve AEP of the responsibility for raising funds to build generation to serve the cooperatives and provide a small profit on the management and backup services the utility would provide. Both companies could benefit from the arrangement.

Buckeye quickly formed a negotiating committee, including Executive Manager Howard Cummins; Buckeye President Owen Manning, manager of Frontier Power Company in Coshocton, Ohio, who later served as president of CFC; Darwin Kindler, manager of South Central Power Company in Lancaster, Ohio; R. Powers Luse, manager of Hancock-Wood Electric Cooperative in North Baltimore, Ohio; and Charles Wyckoff, trustee from Inter-County Electric Cooperative (which later merged with South Central Power Company). They engaged Southern Engineering Company to assist them in evaluating project feasibility and attorneys John W. King and Robert Mone to provide legal support.

Sporn invited Buckeye to participate in the construction of the Cardinal Station. His concept was that Buckeye would own one unit, Ohio Power the other. Plant costs would be shared equally. Ohio Power would provide transmission services to Buckeye's members and purchase any excess generation. There was one condition: Buckeye would have to obtain its funds from the private capital markets, not the REA 2 percent loan program. Thus, Buckeye's decision to access private funds was driven not by the availability of REA funds, although that might well have been a problem had the co-op sought REA funding, but by the condition laid down by Sporn.

Buckeye Engages Katzin and Kuhn, Loeb & Company

Sporn realized that if the project were to be successful, Buckeye would need the help of a knowledgeable investment banker in appraising the economics of the project and in raising funds from the capital markets. Buckeye had established a relationship with the Ohio Power Company in Columbus, but Sporn was concerned that the complexities of the situation required a firm with more experience in utility financing. He introduced the Buckeye leaders to Jerome Katzin, whom he had gotten to know when Katzin served as director of the Division of Public Utilities for the Securities and Exchange Commission (SEC). Katzin became an advisor to Buckeye, and Kuhn, Loeb and Ohio Power agreed to work together to obtain private financing for the cooperative.

At the time, Katzin knew very little about cooperatives or cooperative leaders. "What most impressed me was the fact that these were Ohio Republicans. They weren't wide-eyed New Dealers, they were very conservative people," he said. "Some of them were quite knowledgeable about financial matters, although innocent of Wall Street itself."

The decision to accept Sporn's condition was a difficult one for the Ohio co-ops. "They did not universally agree to it, but he gave them an offer they couldn't refuse," Katzin said.

The Cardinal Station offered the co-ops substantial benefits from economies of scale. Southern Engineering demonstrated through its studies that the cost of energy would be less from a large, privately financed unit than from a smaller unit with 2 percent loan funds from REA.

The total cost of Buckeye's share of the project was approximately $71 million. Buckeye's members contributed about $9 million in equity funds. The remaining $62 million was financed through a private placement at 6.75 percent under the leadership of the Equitable Life Assurance Company. The rate was approximately 0.25 percent higher than the rate then in effect for a double-A credit, the rating held by Ohio Power. "We did not attempt to go public with this issue, because it would have been extremely difficult to explain the cooperatives to the public market," Katzin said.

This was the first time that a rural electric cooperative had raised capital to finance a generating plant in the private markets. "The rate on our private placement seems like a handsome rate today, but in those days it was more than three times what the rural electrics would have paid REA. There was quite a bit of soul searching as to whether they wanted to go ahead with financing on that basis," Katzin said.

Ultimately, they decided to proceed. At the groundbreaking ceremonies for the station on November 4, 1963, Katzin met Clyde Ellis, general manager of NRECA. "He took me aside and asked whether I, as a banker, felt that there would be an opportunity for the rural electric cooperatives, besides Buckeye, to enter the capital markets," Katzin said. "He invited me to come to Washington and meet with him, which I did."

Katzin Comes to NRECA

When Ellis invited Katzin to Washington in 1963, pressures on the traditional REA loan program were continuing to increase. Co-ops were growing at an annual rate of 10 percent or more, almost 50 percent faster than private utilities. Thus, capital demands were increasing at a time when Congress was becoming more and more reluctant to increase loan authorizations. The Kennedy administration, although supportive of Clapp's concepts, was actively trying to reduce the burden on the Treasury. Congressional appropriations for the REA program reached $400 million in 1963, but needs were still increasing. The backlog of loans at REA was over $200 million and projected to reach $300 to $400 million in the near future. Clapp wanted to approve more loans for G&T purposes, but there were not enough funds to meet both the growing needs of the distribution systems and to expand the G&T program. Investor-owned utilities supported massive lobbying efforts to block Congressional appropriations for generation and transmission purposes. REA opponents continued to criticize the 2 percent interest rate. Even the program's friends and supporters in Congress were telling Ellis that the REA loan program had to change.

Ellis presented the problem to NRECA's membership at the regional meetings in the fall of 1963. The program needed adequate access to capital at a low interest rate to meet its future needs, and those needs were expected to increase well beyond then-current levels in the near future. "Let us now determine, with a cold, hard objectivity, where our program stands in relation to the Rural Electrification Act and with the possible avenues of supplemental financing that might be open to us," he said.

In response, members in each of NRECA's ten regions passed

resolutions asking the NRECA Board of Directors to fund a study to examine the feasibility of obtaining funds in private capital markets and to consider whether any changes should be made in the REA financing program. Some of the regions added a qualifying statement supporting the continued availability of the 2 percent program.

After the regional meetings, Administrator Clapp told participants at NRECA's 1964 annual meeting that the REA electric borrowers' needs for capital could total $8.1 billion between 1965 and 1980. "In the light of probable future requirements for new capital in this rapidly expanding electric industry, it certainly is sensible for you to explore the possibilities of supplemental financing in the private money market as your national association proposes to do," he said. Clapp also proposed that REA undertake a study of an intermediate financing program for systems that could afford to pay higher interest rates as an interim step to moving to the private markets.

"What I had in mind at the time was really pretty elementary," Clapp said. "There would be a second loan program in addition to the basic 2 percent direct loan program. The new program would entail a supplemental loan program to provide additional dollars that would be lent on the basis of an interest rate that compensated the government for its cost of money."

Based on the regional meeting resolutions and Clapp's comments, the membership approved Ellis's proposal. NRECA engaged Katzin and Kuhn, Loeb to conduct a study on the possibilities of obtaining supplemental financing from the private markets. Kuhn, Loeb was chosen because of its experience with Buckeye and its success in pioneering the development of new capital plans in various industries.

Ellis told Katzin that he saw the REA program experiencing increasing difficulty with Congress. "These difficulties would become particularly acute as the attention of Congress shifted toward urban problems and away from the farms," Katzin said. "Demographics seemed to make this inevitable. The farms were losing population and losing representation. They no longer carried a strong vote. Looking ahead, he felt that, although cost was very important, availability of funds was even more important if the co-ops were to avoid being starved for capital and eventually put out of business because of their inability to serve their territories."

After the 1964 annual meeting, the scope of the work being done was expanded to include analysis of a broader range of options. NRECA appointed a staff committee to conduct the expanded study and an advisory committee composed of one representative from each NRECA region to review the staff work. Bob Partridge, who later became general manager of NRECA, was a legislative representative at that time. He said it was difficult to get people to agree to serve on the committee. "They knew it was not a very popular idea to even talk about," he said. "Although they did not work in secret, no publicity was given to the work they were doing in trying to develop the framework for a practical approach to supplemental financing. However, there was a very strong feeling among the committee members that supplemental financing was needed."

1965 Financing Study Evaluates Needs

The Rural Electric Financing Study Report was presented to the NRECA board of directors at their July 1965 meeting. The study addressed the key issues that would dominate program discussions for several years. First, it examined program objectives. "From the very beginning, the rural electrification program was set up to serve—and does serve—a social as well as an economic purpose by raising the working and living standards of the people of rural America," the report stated. Thus, area coverage, the concept that any consumer in an area served by a rural electric system should be able to receive service at the same cost and under the same terms and conditions as all other consumers, was an important goal. Territorial protection was an equally important objective, because efforts by private power companies and municipalities to take over populated areas and the more attractive rural loads threatened the ability of many cooperatives to meet area coverage goals at reasonable rates.

The study also supported Administrator Clapp's parity of rates goal to make adequate electric service available at rates that were comparable to those paid by urban consumers in adjacent areas. In 1965, the average monthly use of electricity by cooperative consumers was slightly less than 500 kilowatt-hours. In only four of the 46 states in which cooperatives served were cooperative rates equal to or slightly below the rates of urban utilities at that level of usage. In 35 states co-op rates were more than 10 percent higher than those of urban utilities.

Other program objectives included maintaining financial strength and maintaining the cooperative position in the electric industry. The report also cited the need for cooperatives to be able to benefit from the technological improvements in generation

Under the leadership of General Manager Clyde Ellis (**standing*****), the NRECA board of directors explored other alternatives for supplemental financing. Initially, they tried to get Congress to establish a Federal Bank for Rural Electric Cooperatives before appointing the Long-Range Study Committee to formulate specific recommendations for future financing.***

and transmission and economies of scale, as well as to participate in power pool operations.

The assessment of future capital requirements was an important component of the Rural Electric Financing Study. REA projected capital requirements of $5 billion for distribution and $3.1 billion for generation and transmission for the 1965–1980 period. The Kuhn, Loeb report, which was included as a part of the overall study, made a separate projection of total program requirements ranging from $8.1 to $9.5 billion. As it turned out, those estimates were far too low.

Total utility plant for distribution systems rose from $4.2 billion in 1965 to $13.4 billion in 1980, an increase of $9.2 billion, whereas the G&Ts' total utility plant jumped from $777 million in 1965 to $15 billion in 1980, a $14.2 billion increase. The capital needs during that period were far greater than originally anticipated.

The financing study evaluated several alternatives for meeting capital needs, including the existing rural electric program, the REA intermediate financing proposal, self-financing from revenues, the FCS, guaranteed and insured loan programs, the Buckeye approach, and group action through a nonprofit corporation or public agency.

After extended consideration of these issues, the NRECA board voted to present the findings and conclusions of the study to the fall 1965 regional meetings for action at the 1966 annual meeting. The board recommended that the REA 2 percent interest program remain unchanged for systems needing it, that an intermediate financing plan be developed to provide more flexibility in loans and that a mechanism be established to provide a way to bring supplemental private money into the program.

Electric Bank Proposal

The proposed mechanism was a Federal Bank for Rural Electric Systems that would incorporate the features and experiences of the FCS to provide supplemental capital. The electric bank would operate initially under the supervision of the secretary of the Department of Agriculture using the staff and facilities of the REA. The bank would make loans under terms and conditions similar to those of REA, and loans would be secured on an equal basis with REA loans. The bank would be governed by a board of directors, including federal government and nongovernment representatives. The federal government would provide the initial equity that would enable the bank to borrow funds from private sources. The bank would be authorized to obtain private funds through the sale of debentures, bonds, notes and other debt instruments. It was anticipated that, over time, the rural electric systems would purchase the equity in the bank, and ownership and operation of the electric bank would eventually be transferred to its borrowers.

A copy of the Rural Electric Financing Study Report was sent to each NRECA member well in advance of the 1965 fall regional meetings, so that each system would have time to consider its findings and conclusions. In leading the discussion of the study and recommendations, Ellis emphasized the urgency of the need for change when he said, "Some congressmen who are our long-time friends have plainly told us they would go along with fighting for the 2 percent program for systems that need it, if, on the increased amount, we would either pay the cost of money to the government or borrow the increased amounts from the money markets."

In a message to the same meetings, President Lyndon B. Johnson, long a friend of rural electrification, said, "In the view of many observers, it is reasonable to expect that your requirements for new capital in the years ahead will exceed the amounts that can be provided through the traditional REA loan program. Your decision to explore ways to bring into the program new funds to supplement those available through REA demonstrates once again the judgment and foresight of rural electrification leaders."

To most program leaders, the message was clear. Even rural electrification's strongest supporters would not provide unlimited funds at the 2 percent rate. The cooperatives would have to do something to help themselves.

Opposition from Region 6

A strong minority disagreed. Most of the opposition came from Region 6, which included cooperatives in Minnesota, North Dakota and South Dakota. "There were many people in this part of the country who hadn't bought the idea that we needed a supplemental financing program yet," said Virgil Herriott, general manager of Sioux Valley Empire Electric Association. "We still had a solid basis for expecting to continue receiving 2 percent funding. We believed that if we spent as much time and effort trying to defend the 2 percent program as we did in trying to develop an alternative to it, we could probably prolong its life."

The opposing systems formed the Region 6 Finance Committee, chaired by George Cornog, manager of KEM Electric in

Linton, North Dakota. "Basically, this was a North Dakota–South Dakota committee, organized for the purpose of working together to modify or kill the original NRECA supplemental financing program," Herriott said. "Our deliberations were slanted toward what we could do to make a supplemental financing program unnecessary."

The Region 6 committee engaged its own consultants and focused on regional approaches. "Our thinking at the time was that our views alienated us from the rest of the program," Herriott said. "Our objectives were not the same as the objectives of the people of Georgia, for example. They had an entirely different kind of financing problem from those we were experiencing."

The Region 6 cooperatives had a strong relationship with the federal power program, which helped them to have a low wholesale cost of power. They also had strong support from their Congressional delegation for continued 2 percent financing. "We felt, perhaps falsely, that we were pretty secure out here and that we could follow the traditional way of making our program successful," Herriott said. "We thought there might be a more advantageous way of establishing a supplemental financing source on a regional basis, at least more favorable than getting tied up in a national program. At that time, we perceived that any national organization would have different objectives than we had." The Region 6 committee eventually proposed an alternative plan.

When it became clear that this alternative would not receive adequate support, the Region 6 committee asked for more time for study and discussion. However, the NRECA executive committee was convinced that if it did not submit a plan to Congress soon, the Johnson administration would submit its own.

With the exception of the Region 6 group, the cooperatives generally agreed that a rural electric bank was the most reasonable and economical approach to supplemental financing. When the proposal was presented at NRECA's 1966 annual meeting, the membership voted ten to one to ask Congress to establish the Federal Bank for Rural Electric Systems.

Electric Bank Legislation Fails

Legislation to establish the Federal Bank for Rural Electric Systems and the Federal Bank for Rural Telephone Systems was introduced into the House of Representatives and the Senate in 1966 during the second session of the 89th Congress. Rep. W. R. Poage (D-Tex.), chairman of the House Agriculture Committee, was very supportive of the legislation. Sen. Hubert H. Humphrey (D-Minn.) was a strong supporter in the Senate. The House and Senate Agriculture Committees held hearings, but the session was adjourned before action was taken. The legislation was reintroduced in 1967 in the first session of the 90th Congress. By that time, there was considerable opposition to the bill.

Some of the opposition came from the rural electric systems who opposed supplemental financing. "Any time you have a division among a group that is trying to get something through Congress, you have a problem," said Vincent P. Slatt, manager of Inland Power and Light Company of Spokane, Washington. "We weren't fully united behind the bill. There were those who didn't think we really needed it."

"I viewed it with alarm and dismay. I thought it was a bad idea," Herriott said.

There was even more opposition from investor-owned utilities. "What was most remarkable was the mobilization of the utility industry against this proposal," said Katzin, who continued to advise NRECA. "They had 100 witnesses—100 executives—to oppose the measure. Each of these men took time out from his major utility operation to come to Washington and sit in line to have his say in opposition to this program. The opposition was beyond all our expectations."

The private companies had a well-planned strategy. Opposing the electric bank head-on would have put them in an awkward position. For years they had been making an issue of government subsidies for rural electric systems. It would have been difficult for them to object directly to the cooperatives moving toward financial independence. They were also unlikely to convince Congress that the cooperatives had achieved sufficient financial strength to finance independently. The NRECA and Kuhn, Loeb studies made it clear that the co-ops required special consideration if they were to succeed in approaching the capital markets directly.

Instead, the opponents secured amendments to the legislation that were unacceptable to the rural electrics, focusing on loans for generation and transmission purposes. As long as an investor-owned utility controlled wholesale power supply and the terms and conditions under which it was delivered, distribution systems were attractive as customers and offered little competition to the private power companies. The private companies' real concern was the G&Ts. One amendment, for example, would have given the private utilities the right to challenge any application for REA funds to construct electric generating facilities. If an investor-owned company could show at a hearing before REA, subject to federal court review, that it could supply a cooperative with power cheaper than power the cooperative could generate, the application would be denied. The amendment would have eroded the cooperatives' bargaining position in negotiating power contracts, and co-ops feared this would result in sharp increases in power costs.

"With each succeeding day, some new crippling amendment was shoved into the legislation. By the time it finally was reported out of the House Agriculture Committee and sent to the Rules Committee, it was so badly butchered that we had no alternative but to oppose it," Partridge said. "The bill, as amended, left us vulnerable to court process on a wide range of issues. This vulnerability could have been utilized by anybody who had any objections to rural electrification for whatever reason. We would have been so handicapped in trying to get a loan from REA or the Rural Electric Bank that the legislation was worse than worthless."

The bill was allowed to die in committee on August 1, 1967.

Telephone Bank Legislation Passes

Interestingly enough, the telephone program had a very different experience from the rural electrics. They had strong support from the Bell System and from the U. S. Independent Telephone Association. The telephone systems continued their efforts independently of the electric systems, and legislation establishing the Rural Telephone Bank to provide supple-mental financing for telephone systems was enacted in 1971.

"The difference between the telephone program's experience and that of the rural electrics boiled down to support," Partridge said. "The telephone program had strong industry support, whereas we had strong industry opposition to our bill."

The animosity of investor-owned utilities toward cooperatives was driven largely by economics and philosophy. "The bottom line is always economics," said CFC Loan Officer Richard B. Bulman. "At that time, the unit cost of power production was going down. The more load you could put on and the faster you could grow, the better off you were. The animosity was driven in part also by the fact that the co-ops were borrowing money at a very attractive interest rate." The private companies viewed this as a competitive advantage. The cooperatives were also receiving support from the REA in terms of consulting services, construction design, materials testing and other activities. "If you are competing for the same load but you are not getting the same benefits, then you hate the guy who is," Bulman said.

The private companies understood that they would lose business if the distribution systems developed their own sources of power supply. "They did not like the idea of the co-ops graduating to free markets," Katzin said. "Although historically they could not prevent the REA program from moving forward, the legislative influence of these utilities was sufficiently strong to enable them to restrict the size of the program."

One consequence of the Rural Electric Bank effort was that it focused the attention of the entire rural electric program on the supplemental financing issue. "Because the need for more funding had been so thoroughly discussed among the membership, we thought that there was a consensus on the need for supplemental financing. The attitude of the membership seemed to have changed," Partridge said.

The defeat of the rural electric bank legislation did not affect the cooperatives' need for supplemental financing. It simply changed their approach.

Bank Legislation Failure Leads to Long-Range Study Committee

While NRECA was trying to get more money through the Rural Electric Bank, Administrator Clapp faced a formidable challenge in determining how to allocate the funds available to him. One of his concepts, in keeping with his parity of rates theory, was for the systems whose rates were lower than those of competing investor-owned utilities to receive less than 100 percent of their needs from REA. The idea was that those systems could raise their rates to the level of their competition. That would generate additional cash that could be used to self-finance some construction projects and reduce the demand on REA appropriations.

"We were not satisfied with the parity of rates idea, because the rates in our region were generally much lower than those of adjacent utilities, thanks to the availability of hydro power and 2 percent financing," Herriott said. "We debated Norman Clapp and his staff on the parity of rates concept. We believed

As president of NRECA, Judge T. W. Hunter, attorney for Newberry Electric Cooperative in Newberry, South Carolina, encouraged efforts to address the financing problems facing the rural electric program. Pictured, **left to right:** *Robert D. Partridge, executive vice president and general manager of NRECA; Long-Range Study Committee member Leland Leatherman, attorney for Arkansas Electric Cooperatives, Inc., in Little Rock, Arkansas; and Judge Hunter.*

that consideration should more properly be given to parity of service. We spent one entire afternoon before the NRECA resolutions committee discussing and debating these issues."

At the February 1967 annual meeting, the NRECA membership voted to establish the Parity of Rates and Service Committee. The ten-member committee was asked to study the parity issue and also to consider the possibility of having a multi-interest rate loan program at REA. J. K. Smith was appointed chairman. Herriott represented Region 6 on the committee.

Inland Power's Slatt, the Pacific Northwest's representative on the committee, said that by the end of the first meeting there were about 35 suggested criteria for establishing parity. Co-ops in areas with high labor costs wanted labor costs to be a factor. Co-ops in Michigan that were paying as much 15 mills for power, compared to the 2.5- to 3.0-mill power cost in Slatt's area, said power costs should be a factor. States that paid property and gross receipts taxes said taxes were the issue. After two or three meetings the committee reached the conclusion that it couldn't come up with a set of criteria it could recommend to REA.

Judge T. W. Hunter, attorney for Newberry Electric Cooperative in Newberry, South Carolina, was president of NRECA at the time. "We told Judge Hunter, 'The problem isn't parity of rates. The problem is, we don't have enough money,' " Slatt said. "Judge Hunter wrote back to J. K. Smith and Clyde Ellis that if parity of rates isn't the problem and financing is, then you had better settle the problem of financing."

After the failure of the rural electric bank legislation, the NRECA board faced two key questions: Where does the rural electric program go from here, and how does it get there? The board restructured the Parity of Rates and Service Committee to form the Long-Range Study Committee in August 1967 and gave it the responsibility of considering all factors involved in formulating specific recommendations concerning financing. Most of the members of the Parity of Rates and Service Committee, including Herriott and Slatt, continued to serve on the 26-member Long-Range Study Committee. Smith again agreed to serve as chairman.

First Large-Scale, Program-Wide Study

The formation of the Long-Range Study Committee marked the first time the rural electrics had addressed a major issue on such a large, program-wide scale. Its methodology provided a template used by subsequent committees studying later issues affecting the entire program. The committee members were carefully selected to reflect the diversity of the program. Each region of the country was represented. Each segment of the program—distribution systems, G&Ts and statewide

The members of the Long-Range Study Committee reflected the diversity of the rural electric programs.

J. K. Smith
Chairman
Manager, Kentucky Rural Electric Cooperative Corporation
Louisville, Kentucky

Virgil Herriott
Chairman of Objectives Subcommittee
Manager, Sioux Valley Empire Electric Association
Colman, South Dakota

Vincent P. Slatt
Chairman of Financing Subcommittee
Manager, Inland Power and Light Company
Spokane, Washington

J. R. Cobb
Chairman of Implementation Subcommittee
Manager, Texas Electric Cooperatives
Austin, Texas

Ernest Ballard, Jr.
Manager, Moon Lake Electric Association, Inc.
Vernal, Utah

C. E. Boulson
Manager, Sho-Me Power Corporation
Marshfield, Missouri

James Campbell
Manager, Poudre Valley Rural Electrification Administration
Fort Collins, Colorado

W. C. Carlton
Manager, Carteret-Craven Electric Membership Corporation
Morehead City, North Carolina

John R. Dolinger
Manager, Cumberland Electric Membership Corporation
Clarkesville, Tennessee

James T. Dudley
Manager, Singing River Electric Power Association
Lucedale, Mississippi

Clyde T. Ellis
General Manager Emeritus, National Rural Electric Cooperative Association
Washington, D. C.

Sailey Ennis
Manager, Washington Electric Cooperative, Inc.
East Montpelier, Vermont

Walter Harrison
Manager, Georgia Electric Membership Corporation
Millen, Georgia

Earl L. King
Manager, Iowa Association of Electric Cooperatives
Des Moines, Iowa

Leland Kitt
Treasurer, Southwest Public Power District
Palisade, Nebraska

Leland Leatherman
Attorney, Arkansas Electric Cooperatives, Inc.
Little Rock, Arkansas

E. V. Lewis
Manager, Central Electric Power Cooperative, Inc.
Cayce, South Carolina

T. C. Long, Jr.
General Manager, Walton Electric Membership Corporation
Monroe, Georgia

John P. Madgett
Manager, Dairyland Power Cooperative
La Crosse, Wisconsin

Owen Manning
Manager, The Frontier Power Company
Coshocton, Ohio

R. B. Moore
Manager, Lea County Electric Cooperative, Inc.
Lovington, New Mexico

Lloyd Patton
Director, Cotton Electric Cooperative
Walters, Oklahoma

Clarence W. Peterson
Manager, Lake Region Co-op Electrical Association
Pelican Rapids, Minnesota

Harlan E. Ruback
Manager, Top O'Michigan Rural Electric Cooperative
Boyne City, Michigan

Eugene E. Taylor
Manager, Concordia Electric Cooperative, Inc.
Ferriday, Louisiana

Cecil E. Viverette
Manager, Blue Ridge Electric Membership Corporation
Lenoir, North Carolina

T. W. Hunter, Ex Officio
President, National Rural Electric Cooperative Association
Attorney, Newberry Electric Cooperative
Newberry, South Carolina

organizations—was represented. Both large and small cooperatives had a voice. Each committee member agreed to serve without compensation and at his own expense or at the expense of the organization he represented. Perhaps most importantly, the individual committee members were selected on the basis of demonstrated leadership ability and professional qualifications and experience. The Long-Range Study Committee was composed of the best and brightest the rural electric program had to offer.

Another key element was the committee's decision to fully involve the membership in its activities and deliberations. It did not act alone. The committee met almost every month from November 1967 to the conclusion of its work in March 1969. It kept complete minutes on all its activities and regularly distributed reports through NRECA publications, special memoranda and presentations at statewide and other general meetings.

The committee held open-forum hearings in six cities during the spring and summer of 1968 to solicit the views of the membership. After the bulk of its work was completed, it held a series of "fly-in" meetings in five cities during the week of January 27, 1969, to hear final comments and concerns. The committee listened to any idea or viewpoint offered.

"One of the things that I liked most about working with the co-ops was how democratic it was," Katzin said. "I saw for the first time participation by people in decisions. On Wall Street, everything was very autocratic. The boss told you what to do, and you did what you were told. Suddenly, I was thrust into this group who had long-range study committees, listened to everybody and explained everything. To me, that was a different order of things."

Hundreds of rural electric representatives throughout the nation contributed to the committee's efforts. Even those who opposed supplemental financing aided the effort by highlighting weak points in the committee's analysis. Their insights strengthened the plan ultimately presented to the members.

No Preconceived Ideas

The committee held its first formal meeting on November 13, 1967. "I was most impressed with the committee's insistence on embarking on its assignment without any preconceived notions as to what its final recommendations will be," Smith reported to the NRECA board and statewide organizations after that meeting. "We heard many far-ranging challenges and ideas from those who appeared before us this week. I'm sure we will hear others during the coming year. All of this makes us realize we can't limit our thinking."

In reflecting on that meeting some years later, Smith said, "We were going to approach the study with open minds and with the understanding that ideas would have to come from the membership. Not one man on the committee had any idea concerning what would eventually evolve, what structure we were going to establish or what type of financing we were going to recommend. The idea of CFC as a supplemental financing organization came from discussions and exchange of ideas with our members as well as from the committee's consultations with financial experts. You can't attribute the final recommendations to any one individual or group of individuals."

Chairman Smith told the group it faced three tasks:

- to take a new look at the objectives of the rural electric cooperatives at a time of great change in the nation and the world,

- to develop recommendations on alternative approaches to satisfy the long-range financial needs of both the distribution and power supply systems and

- to implement necessary changes in the program with the support and assistance of the cooperative systems, the general public, Congress and the executive branch of the federal government.

The committee established three subcommittees to address these issues. Region 6's Herriott was selected as chairman of the Objectives Subcommittee. Its job was to develop and define long-range objectives for the rural electric systems. Inland Power's Slatt was appointed chairman of the Financing Subcommittee. Its assignment was to develop a financing plan that would meet total program needs and be compatible with program objectives. J. R. Cobb, manager of Texas Electric Cooperatives, Inc., chaired the Implementation Subcommittee. That group was to develop plans for reporting to the membership, involving them in committee deliberations and building support for and under-standing of whatever recommendations developed.

Kuhn, Loeb's Katzin agreed to advise the committee on financial issues. The committee tapped former REA Legislative Consultant Charles Samenow as legal advisor. Several members of the NRECA staff, including Kermit Overby, Ira Shesser, Dick Pence, Phil Sawicki, Gary Tabak and June Kraeft, were assigned to work with the committee. Clyde Ellis, by then general manager emeritus of NRECA, was also an advisor to the committee.

The leaders of the Long-Range Study Committee insisted that the committee approach its work with an open mind and actively solicited input from all segments of the rural electric program. Pictured, **left to right:** *Long-Range Study Committee Chairman J.K. Smith, Objectives Subcommittee Chairman Virgil Herriott, Financing Subcommittee Chairman Vincent P. Slatt and Implementation Subcommittee Chairman J. R. Cobb.*

Katzin credits Ellis, who had never met a Wall Street investment banker before talking to Katzin at the Cardinal Station dedication, with the vision that led to the creation of CFC. "He had the vision, he had the drive, and he had the political sense to get it off the ground," he said. "Some people say that history has its own destiny and it doesn't depend on individuals, but certainly it depended on him at that time. There would have been no CFC without him."

Rep. Poage spoke to the Long-Range Study Committee at its first meeting about the outlook for a legislative solution and the unsuccessful attempt to pass supplemental financing legislation in the 89th and 90th Congresses. "You and I both underestimated Congressional support for new rural electrification program legislation," he said. He urged the systems to give unanimous support to whatever program they structured and to make sure that the general public and members of Congress understood the accomplishments of the rural electrification program and the need for additional financing. He also advised that no further legislative attempts be made during the 90th Congress and possibly not in the 91st.

The committee established two target dates. It wanted to present an interim report to the members at the regional meetings in the fall of 1968, and it wanted to present a final report for member action at NRECA's 1969 annual meeting.

The committee worked hard to meet that schedule. "I got mad at J. K. Smith because he always called his meetings for seven o'clock in the morning," Slatt said. "Well, seven o'clock Washington, D.C., time was four o'clock Spokane time, but J. K. was one of those people who wanted to start on time and go hard all day long."

The committee attracted high-level attention both inside and outside the rural electric program. A highlight of the December 1967 meeting was a surprise visit from Vice President Hubert H. Humphrey. Federal Power Commission Chairman Lee C. White, who was speaking at the dinner session, quickly yielded to the vice president. "I heard you fellows were in town, so I thought I would drop in for a visit," Humphrey said. White never got to finish his remarks, but everybody there appreciated Humphrey's demonstration of his interest in helping to find additional sources of capital for rural electrification.

Survey Solicits Viewpoints

On January 11, 1968, the Long-Range Study Committee mailed a questionnaire to the directors, managers and attorneys of all rural electric cooperatives in the nation. It was designed to solicit program viewpoints and concerns on a broad range of issues. The committee asked each system to complete the questionnaire at a board meeting, so that everyone would have an opportunity for input. If a system returned the questionnaire without comment or answers, the committee asked the cooperative to submit its general views about future objectives and financing.

By the April 12, 1968, due date, 83 percent of the systems had responded. Ultimately, 873 systems, or 90 percent, returned a completed questionnaire. The results reflected the members concerns about the power supply and financial situation at the time.

Two key questions established the attitude toward the need for supplemental financing. One was whether rural electric systems needed a source of capital other than general funds to meet long-term needs. Over 97 percent of the systems responding by the April 12 deadline said yes. The second question asked if the system believed it could depend solely on government financing as that source. More than 80 percent said no.

Another important question was whether the system considered the cost of money or the availability of money to be more important. Without access to adequate funding, the systems could not continue to exist. On the other hand, outside capital was obviously going to cost more than traditional REA loans. "The survey established that the consensus thinking was that availability was more important than cost," Smith said. "This gave us a good foundation on which to begin our work."

It was a tough choice. "A number of the systems noted that they believed each of the factors, cost and availability, was of equal value," the survey report said. "However, it is significant that the overwhelming majority recognized that if money was not available, it was of little importance what it cost."

The survey respondents ranked providing reliable service as the most important of five service issues and providing dependable power supply on reasonable terms second. Low retail rates were ranked as the least important. "It would appear very likely that the systems are more concerned about maintaining competitive rates rather than low rates," the survey report said.

Regulation and territorial protection would be major concerns for the private markets. Over half of the systems reported territorial protection in some form. Almost 80 percent of the nonregulated systems said they would accept regulation in return for territorial protection. Of the 372 systems that were already regulated, only 12.4 percent reported unsatisfactory experiences. The systems also indicated a strong preference for the cooperative form of business structure to carry out rural electric needs, ranking it ahead of nonprofit stock corporations, profit corporations and public utility districts.

The survey revealed a consensus, if not unity, in current member thinking on three issues that have, in many respects, dominated CFC's history: availability of capital, cost of capital and ownership. It gave the committee a sound basis for its work.

Members Speak at Open Forums

The committee solicited additional member input at a series of two-day open-forum meetings scheduled over a six-month period in Atlanta, Salt Lake City, St. Louis, Sioux Falls, Dallas and Washington, D.C. The committee reported on its current activities, then gave the membership the opportunity to make oral presentations and file written statements. The forums were attended by a total of 1,400 people. One hundred fifty-nine rural electric leaders spoke about their concerns and ideas. Each statement was reviewed by the appropriate subcommittee, which then recommended action to the full committee.

Although the main issue was financing, other program concerns were discussed as well. Territorial protection and power supply issues were major concerns. Many witnesses supported the need for cooperatives to develop their own power supply sources.

The systems expressed their desire to remain nonprofit, consumer-oriented cooperative organizations. Their primary objective, they said, was still to provide electricity reliably, whenever and wherever it was needed, at the lowest possible cost. Many also stressed the need for cooperatives to become involved in rural area development, but there was a range of views on just what that meant. Quite a few systems urged other cooperatives to consider the advantages of merging or sharing services.

As the work of the committee progressed, the comments and suggestions became more focused. Early on, a consensus emerged that Congress was unlikely to appropriate sufficient REA funds in future years to meet the rural electrics' requirements. "We do not feel that we can continue to look to the Rural Electrification Administration to meet our future capital financing needs," said Barbara H. Deverick of Blue Ridge Electric Membership Corporation at the March 12 Atlanta meeting. "Nor do we believe, with the going market rate for money, that we can continue to finance at 2 percent." Although there was strong support for maintaining the 2 percent program for co-ops that needed it, many cooperatives supported a market-rate program for the stronger systems.

There was also an emphasis on the need for program unity. "The one force that has held us together for these 30 years and enabled us to achieve some degree of success is our common banker, REA. REA has given the rural electrification program the national unity of purpose that is so necessary for success," said Colorado-Ute General Manager John Bugas at the April 8 Salt Lake City meeting. "If we are to continue with the same national unity, we must set up our supplemental financing on a national basis, and not on a regional, state or individual cooperative basis as some have urged. To go into the money market individually will surely lead to our downfall."

Committee Explores Many Options

Early in their deliberations, the Long-Range Study Committee explored the possibility of obtaining supplemental financing through the FCS, an umbrella organization for specialized banks and credit associations. The FCS had been quite successful in raising capital through the sale of securities in the private market. Many rural electric consumer-members had a credit relationship with an FCS organization. It provided one-fifth of the mortgage funds used by farmers, 13 percent of operating credit used by farmers and 60 percent of all loans to farm cooperatives. One division, the Banks for Cooperatives, had been established specifically to make loans to cooperatives.

The committee saw that a relationship with the FCS might offer mutual benefits. Loans to electric cooperatives would help the Banks for Cooperatives broaden their asset base and increase their leveraging capability. The rural electrics would get access to the loan funds they needed from a familiar, if new, source. The committee decided to approach the FCS about developing a supplemental financing program for electric cooperatives.

R. B. Tootell, a graduate of Washington State University, was governor of the FCS at that time. Hoping to build on common ground, the committee asked Slatt, whose co-op was located in Washington, to head the delegation to approach Tootell.

"We went to see Governor Tootell, and he told us in no uncertain terms that he wasn't interested in the problems of the rural electrics. He stated flatly that he wasn't interested in working with us," Slatt said. "We were very disappointed because we thought the Banks for Cooperatives would be the ideal place for us to obtain the supplemental capital we needed."

Tootell was invited to address the issue with the full committee at its January 1968 meeting. Tootell explained the organization of the FCS, then gave two reasons why he did not support an expansion of FCS lending programs into the rural electric program. In the fall of 1966, the Johnson administration, struggling with the Vietnam War and the budget deficit caused by expenditures for the war, asked the FCS to restrict its borrowings from the private capital markets. "We feel that we will be actually pushed to get all the loan funds needed by the types of borrowers to which we now have a responsibility," he said. "We are somewhat reluctant to take on another responsibility."

Beyond that, Tootell said the FCS did not want to become involved in a program that some considered controversial. "Why should the Farm Credit System get into a program like rural electric financing that would involve FCS in substantial controversy in the political arena?" he asked.

"He told us that they had considered the kind of adversary relationship that we had had with the investor-owned utilities over the years and that they didn't think that they wanted to include within their membership a group that was going to bring with it a public battle," Herriott said. "They took pride in their ability to be able to borrow in the marketplace at very favorable rates and in the good working relationships they enjoyed with finance organizations in the money centers of the United States. They knew that those same institutions and organizations had friendly working relationships with the investor-owned utilities."

There is a certain irony to Tootell's position in light of subsequent events. By 1972, the FCS was actively soliciting the rural electric business in direct competition with CFC. But if Tootell had said yes, there would never have been a CFC.

Tootell suggested that the electric cooperatives set up an organization patterned after the FCS but the rural electric bank legislation, which would have done that, had already failed twice. The committee's appraisal of prospects in the 91st Congress ultimately led it to recommend against pursuing that option.

Financing Subcommittee Seeks Ideas

The Financing Subcommittee held a series of dinner meetings in conjunction with the open-forum meetings to discuss various financing alternatives with financial experts in different fields. The range of topics covered illustrates the thoroughness of the subcommittee's investigations. The full committee was invited to participate in the meetings as well as the staff and special consultants.

The first of these took place on the evening of March 10, 1968, in Atlanta, Georgia. Joseph C. Swidler, who served as general counsel to the Tennessee Valley Authority (TVA), chair of the Federal Power Commission and chair of the New York State Public Service Commission, suggested a concept for issuing tax-exempt securities. His idea was for state legislatures to authorize rural electrics to issue tax-exempt bonds, with federal legislation also a possibility. REA would agree to subordinate its debt, that is, agree that its debt would be paid after the tax-exempt bonds in the event of a bankruptcy or foreclosure, to facilitate the transaction. The advantages to the government, he said, were that its security would not be impaired, there would be no new special advantage (the 2 percent loans were already tax-exempt) and the rural electric loan program would be removed from the federal budget. From the co-op perspective, they would no longer need to go to Congress for appropriations each year, and it would open up capital markets not previously available to the rural electric systems.

At the April meeting in Salt Lake City, the subcommittee heard from J. Orin Shipe, managing director of CUNA International, Inc. He reviewed the history of the credit union movement, its international aspects and his organization's relationship with the rural electric program. He offered to work with the committee on establishing a national credit union and also suggested that if the committee decided to use a nongovernmental approach, it should consider a profit corporation. "Rural electric cooperatives may find that there is greater flexibility in organizing under the corporation statues," he said.

The committee also heard from Lawrence Leighton, an associate with Kuhn, Loeb, who suggested that five basic criteria should be considered in developing a financial policy: viability, capital at least cost, flexibility, control over the destiny of the program and maximum leverage. "You should give equal weight to all of these items," he said. "It would be unwise to put too much emphasis on any one item."

The next Financing Subcommittee meeting, which was attended by all members, was held during the St. Louis open forum on the evening of May 13, 1968. The meeting turned out to be a pivotal point for the rural electric program. The first speaker was Bernard Goldhammer, assistant to the administrator for marketing and power for Bonneville Power Administration. He told the group about Bonneville's contract financing approach, which used power contracts with municipals and public power districts as security for money borrowed to finance power plant construction. "Because the revenue bonds of municipalities and other public bodies are tax-exempt, this contract financing has enabled borrowings at some of the lowest rates of interest," he said. Under existing laws, however, the cooperatives were not authorized to participate in tax-exempt financings.

Katzin's Idea

Kuhn, Loeb's Katzin was the next speaker. He presented a memorandum proposing a new type of financial institution. It would not be a commercial bank. "It would not meet your needs at all if it were subject to the local or national banking laws and under the direct regulation of a state commissioner of banking," he said. Neither would it be an insurance company. "Only a small portion of the reserves of such a[n insurance] company would be available for lending to the co-ops for any particular project, because insurance companies are closely regulated, too, and there are strict rules governing investments and requiring broad diversification of risk."

Jerome S. Katzin of Kuhn, Loeb & Company first presented his concept for the organization that became CFC to the members of the Long-Range Study Committee during the open forum in St. Louis in May 1968.

The organization would be similar to a mortgage company or a finance company, but its loans would be made only to member systems. "Based on its equity capital and the quality of its long-term loan portfolio, it would itself borrow funds in the capital markets. These borrowings would be both long and short term, including, when appropriate, the issuance of commercial paper," Katzin said. "How much the institution could borrow, how large it could become, is impossible to state at this time. This would depend on the size of its equity capital; on the nature, maturity and quality of the loans it makes to the co-ops; on its earnings from such loans; and on the reputation that it acquires over a period of time for prudent and able management of its affairs."

Katzin said it would be useful but not essential for the obligations of the institution to be guaranteed by the federal government or insured in some fashion. However, that would require federal legislation.

For the institution to succeed, it would need a sizeable amount of equity capital. "Failing federal appropriation, where can it come from? I suggest that it can come from yourselves, from your own bank account," Katzin said. "I am referring to the $300 million or more of prepayments that you have made on your REA loans. This is a sizable sum and enough to capitalize a very respectable financial institution. It would rank among the largest in the country with such a capital base."

Katzin proposed that the systems stop making prepayments to REA until loans were current, which would probably be a two- to three-year period. The payments that would otherwise be made would be used instead to subscribe to the capital of the new financing institution. With $300 million in equity, he estimated that the institution could borrow between $300 and $600 million for total resources of $600 to $900 million.

The organization could continue to increase its equity by requiring additional equity subscriptions from borrowers for each loan, by selling an intermediate security to the co-ops and their members, by placing a surcharge on billings and by accumulating earnings on its loans. (Not all of these approaches were ultimately implemented.)

Aside from the incentive provided by the need for an additional source of capital, the institution would pay a return on the equity investment in excess of the 2 percent the prepayments were earning. The plan would require the cooperation of REA in several key areas, none of which appeared to require legislation.

"I believe that our experience and studies have now reached the point where they lead to one conclusion: that supplemental financing must be derived solely through self-help without reliance on any legislation," Katzin told the committee. "The organization and financing of this lending institution through your own efforts and under your control should not impair the continuation of the present government loan policy for those projects that continue to require lower interest rates and that meet the standards of the REA."

Forming a Consensus

The committee was so enthusiastic about the idea that it asked Katzin to present the paper to the members at the meeting the next day.

"He stated that if we could get our own equity, he was reasonably sure that we could get Wall Street to come along. To me, the idea that if we had our own equity we could successfully obtain outside financing was the trigger point when the decision was made that we would have a supplemental financing program," Slatt said. "That approach was much different from asking the government to fund a program."

Although they would not make a final recommendation for some months, the committee began to form a consensus to follow the self-help path. "From that germ of an idea came the much more complex and sophisticated form of institution that we know today as CFC," Katzin said.

The committee continued to listen to and explore other ideas. Floyd Lamb, investment officer with John Hancock Mutual Life Insurance Company, addressed the Financing Subcommittee meeting during the June 1968 meeting. He described the key elements of financing with institutional investors, such as insurance companies. He said that his company would not be interested in financing very much of the rural electric program, because they needed growth and higher yield securities than it would be able to offer.

"We just couldn't find any connection or any way whatsoever to go the insurance route. It just didn't pan out," Slatt said.

The Financing Subcommittee met with G. O. Wessenauer, manager of power for the TVA, during the July open forum in Dallas. Wessenauer described TVA's approach to legislation that enabled them to go to the open market for financing for power needs. The key aspects included a limitation on service area; approval by the Treasury secretary of the issuance of any securities, including interest rates; and self-liquidation of all loans for power development.

The subcommittee also heard from John Bockoven, president of Federated Electric Insurance Corporation. He said that a study prepared by Nationwide Insurance Company indicated that if farmers served by rural electric systems purchased all of their insurance needs from one company, there would be enough capital available to finance the rural electric program.

Dr. Paul Nelson, clerk and staff director of the House Committee on Banking and Currency, met with the Financing Subcommittee during the August open forum in Washington, D.C. He identified three different approaches the cooperatives could use to solve their financing problems. The first would be to continue to obtain REA loan funds but at a higher rate of interest. The second would be a completely private approach, similar to the one proposed by Katzin. The third would be a blend of federal and private sources. "Electric cooperatives should face the subsidy issue squarely," he said. "Any subsidy should be justified on the basis of national interest involved, such as the immense task of rural area development." He encouraged the cooperatives to continue their relationship with the federal government in some form.

The subcommittee heard about the criteria for obtaining funds through state investments from Robert Blixt of the Minnesota State Investment Board. Blixt said the state public employee retirement funds totaled about $40 billion, some of which could be lent to cooperatives. Most states, however, required utilities to have a triple-A credit rating, a standard most co-ops could not meet. The states might be able to invest in a private credit institution, particularly if it were backed by the government in some way. He said that state fund investors avoided investing in controversial programs. "State investors would need evidence of cooperation between investor-owned utilities and rural electric cooperatives before they committed any funds," he said.

Co-ops Make Suggestions

During this period, the committee also received many suggestions from within the program concerning various financing alternatives. The Ohio systems suggested a rural electric trust fund capitalized through a five-year surcharge on kilowatt-hour sales. Panhandle Rural Electric Membership Association suggested an electric credit bank that would buy bonds issued by rural electric systems and sell debentures on the open capital markets. Many systems favored working through an insurance company.

Several systems supported the concept of government guarantees of the obligations of individual systems. Ultimately, the committee found this to be impractical because it did not provide an assured source of capital and it involved the highest interest cost of any of the alternatives considered.

Another popular suggestion was to obtain authority for cooperatives to issue tax-exempt securities. The committee did not believe that option could succeed, given efforts by the Treasury Department and Congress to curtail the issuance of tax-exempt municipal industrial revenue bonds. Again, there was a strong desire to find a solution that did not require legislative action.

The systems were clearly giving serious thought to the need

for a dependable source of low-cost supplemental financing. There were also those who continued to believe that getting additional funds for the REA program was the most feasible approach.

"Until we have something else that will do the job, we must depend on and fight for these funds. We must make greater efforts than we have in the past," said Virgil T. Hanlon, general manager of East River Power Cooperative in Madison, South Dakota. "We do not have to be ashamed of the success of the greatest program in the history of rural America. At this moment, we have no other means of financing."

Gary M. Williamson, manager of Central Power Electric Cooperative, Inc., in Minot, North Dakota, agreed. "It is most difficult for me to understand that some involved in the REA struggle can advance the argument that the cooperatives willingly accept that the 2 percent loan program is and should be a thing of the past. Such an optimistic viewpoint is hardly consistent with what we know to be our own needs and problems," he said, citing a need for transmission facilities in his area. "It would be impossible for our cooperative or any of our member cooperatives to meet those needs if the 2 percent loan program were abolished or drastically changed."

With the first deadline for an interim report at NRECA's fall 1968 regional meetings rapidly approaching, the committee had to work quickly after the conclusion of the August open forum to finalize its presentation. The committee debated the advantages and disadvantages of the various approaches. NRECA General Manager Bob Partridge told the committee that, no matter how it was structured, the picture was not bright for getting legislation. "I do not believe any major rural electric bill could go through the Congress without serious amendments adding severe restrictions," he said.

An independent credit institution would give the systems immediate control and ownership and provide an adequate source of capital in the long run, but it would take time to develop, its cost would be higher and the systems would have to invest a substantial amount of capital in the organization.

The committee decided to make no recommendation on financing at the regional meetings. Its report would be for information and discussion only. It would request no action at that time.

The interim report projected that estimated capital investments in rural electrification would range from $8 to $12 billion over the next 15 years. If Congress continued REA loan authorizations at the current level of $350 million annually, a minimum of $3 billion would be needed from sources other than REA. If the systems continued to grow at the present rate, almost twice the $5.5 billion invested in the previous 33 years would be required in half the time.

The committee offered three approaches for obtaining the additional financing:

- continued dependence on the existing REA 2 percent loan program to meet all needs,
- supplemental financing obtained through legislation and
- supplemental financing via the self–help route.

It waited for the members to express their views.

Rural Electric Viewpoints and Objectives

While the Financing Subcommittee was exploring ways to raise money, the Objectives Subcommittee was giving serious thought to what it meant to be a rural electric cooperative. "The committee spent almost as much time on developing viewpoints and objectives as it did on the structuring of a financing vehicle, because it was believed that the heart of the rural electric program was more than just providing electric service," Ira Shesser said. The rural electric systems were usually the largest business located in their service area. "The Long-Range Study Committee felt it was very important for the rural electric systems to assume leadership roles and responsibility for helping to improve not only the economic life of the areas they served but also to assist in the development of much needed social services. They recognized that the rural areas required development and that the rural electric systems could serve as the nucleus for generating the necessary support for programs to undertake this type of activity. The members of the committee also believed that such collateral activities were important in developing further justification for the rural electric program and assuring that it would have the strength and support in Congress that was needed."

Developing a statement of viewpoints and program objectives

was an integral part of the scope of the committee's work. The viewpoints formally stated the concerns and philosophies of the program as a whole. The objectives were broad guidelines defining what the program needed to do to support the viewpoints. The committee also suggested activities, specific work different segments of the program could do, in support of the objectives. "We needed formal objectives to define what we were attempting to finance," Smith said. "Objectives addressed

The Day the Lights Came On *by Clyde Denton*

At a time when most consumers view an electric cooperative as just another power company, it is often hard to explain why the rural electric program is so special. People who can't remember a time when there was no television don't always appreciate how wonderful a few light bulbs and an electric iron were to those who had labored without them. In this reminiscence, former CFC Information Officer Clyde Denton shares the excitement his family felt when electricity came to rural Kentucky.

Electricity came to the house I lived in as a boy in the spring of 1939. It was a red-letter day for my family. With the nation still in the throes of the Great Depression, there weren't many days we could look back on with much satisfaction or look forward to with eager anticipation. We had been watching the rural electric lines make their way through the Kentucky countryside toward our farm for months. Progress seemed agonizingly slow, and we had little hope that this modern miracle would ever become a reality for our household.

The first time my family really believed that one day we would have electricity was when my uncle, who owned the small farm where we lived and that we tended, granted the local rural electric cooperative a right-of-way for electric lines. Like most of his neighbors, he was so happy to be a candidate for electric service that he didn't even think of asking for payment.

Of course, we had a few neighbors in our area who didn't understand electricity and refused to let the lines come through their property. They believed the tales being told, for example, that electricity could leak out of the lines and cause the cows to quit giving milk and the chickens to stop laying eggs and, even worse, that it could induce sterility in people living near the lines. In such cases it was necessary for the rural electric cooperative to "dog leg" the lines around the protesting farmer's property.

To my family, however, electricity meant that my father and I would no longer have to walk the four miles to the nearest gas station to hear President Roosevelt's fireside chats or a sportscaster from Madison Square Garden in New York describe the prizefighter Joe Louis's great feats in the ring. It meant that my two brothers, my sister and I would no longer have to crowd around the kitchen table at night trying to do our homework in the feeble, yellow light of a kerosene lamp or to dash to the door and throw the lamp into the yard to keep it from exploding when the flames traveled down the lamp wick into the reservoir of kerosene.

My mother saved for months to pay the $4 membership fee to our rural electric cooperative, numerous questions that had to be considered because of their direct impact on our future financing needs. For example, it was our viewpoint that the rural electric program preferred to maintain the cooperative approach in its corporate structure. If that was our viewpoint, we needed to establish objectives to implement it."

Objectives Subcommittee Chairman Herriott has acknowledged that he was initially opposed to the supplemental

thus assuring us of electricity if and when the lines reached our neighborhood. It was a tough job, because my father was unemployed, and we were subsisting on what we could raise on a small acreage of relatively unproductive soil. Many a neighbor paid the membership fee by using chickens and eggs as barter.

With the advice of the electric cooperative, my uncle and father wired each of the three downstairs rooms in our house for a single light bulb and put an extra outlet in the kitchen. There was no wiring for the upstairs, which was more of an attic than a room and was used primarily as a bedroom for the children.

Once the rural electric cooperative gave us a date on which we could expect electricity to come to our house, we were beside ourselves. When the big day finally arrived, we turned the switch on first thing in the morning. Then we waited for the electricity to arrive. All other activities for the day were canceled.

Late in the afternoon the three light bulbs flickered. Then they lit up. There was no cheering or applause. We simply sat and stared at the glow as if in a trance. Even after the silence was broken, we sat up most of the night watching the light bulbs burn. It is the only time I can remember when my father allowed us to be so extravagant with the use of electricity.

Before rural electrification, my mother did the family laundry by bending over a scrub board to wash the clothes and using four "sad irons," heated over a wood stove, to iron them. I still remember her knuckles that were often raw and bleeding and the burns from the irons that scarred her arms. Electricity added years to my mother's life. It gave our entire family great happiness and freed us from much drudgery. It made a more complete life possible for rural America.

financing concept. "We [Region 6] never viewed the Long-Range Study Committee as being primarily a finance committee," he said. "What we saw in the Long-Range Study Committee was a group that would define a uniform set of objectives for the program. Our theory was that if we could get that job done, then the rest of the program would be able to see that embarking on a silly supplemental financing program was not reasonable. We believed that until Congress turned us down on federal funding, it was ridiculous for us to suggest to anybody that we develop some kind of alternative to federal funding."

There were various views among the membership as to what a cooperative's role in the community should be. "It became clear to me, as we began to debate financing alter-natives in the middle 1960s, that there was a difference in objectives," Herriott said. "Some people had a very narrow view about what their objectives were. They were leaving out what we [the subcommittee] thought were some very important areas." The subcommittee tried to define what those areas were. "Our major concern was about what we needed to do as organizations that were beginning to mature and that had facilities in operation."

In addition to the issues related to providing electric service, the subcommittee identified key areas of performance, such as corporate citizenship, management and leadership, political action and use of natural resources, that were also important to any other type of business. "We thought that objectives ought to be set in each of these key performance areas," Herriott said.

The objectives generated a great deal of comment and discussion at the open-forum meetings. One widely discussed issue was whether cooperatives should try to be member services organizations. "This question concerned whether we should have such services as wiring inspections and merchandising," he said. "It was questioned whether the cooperative should have any kind of advice and assistance programs. Many cooperatives had electrification advisors on staff to work with the membership to teach them how to use electricity efficiently and economically. There were also many cooperatives that didn't believe in that type of service. Others felt it was an important area of responsibility."

The subcommittee found the member survey to be helpful. "By asking the right questions in our surveys, we could cause people to think about things we considered important," Herriott said. "Eventually we were able to develop a sense of awareness and of support for the idea of adopting program objectives."

From the beginning the purpose of the rural electric program was to provide electricity to people and businesses in rural America. The work of the Long-Range Study Committee went beyond that to address in a comprehensive manner how that should be done in the present and future. The committee's 16-page statement of viewpoints and objectives affirmed the programs support for such diverse issues as nonprofit operation and consumer ownership, natural resource development, increased power usage, territorial protection, community development and corporate citizenship responsibility.

Consensus on the viewpoints and objectives did not derail the supplemental financing efforts: it supported the need for it. "As I got more deeply involved in the financing studies and as I was exposed more and more to the arguments of people with an opposing point of view, I began eventually to change my mind on some things," Herriott said. "I came around to the point where I could say that I would support a supplemental financing program for those who needed it, but that I would oppose that program being imposed on the rest of us. I acknowledged the fact that, based on all reasonable projections, there was no way in which we could get enough capital from REA to meet our future financing needs."

The philosophy that emerged from the efforts of Herriott and his subcommittee served as a guide for the rural electric program for many, many years.

Regional Meetings Examine Options

Each cooperative received a copy of the Long-Range Study Committee's interim report in advance of the fall 1968 regional meetings. Half a day at each meeting was devoted to discussion of the committee's work. Each subcommittee presented its report. The committee members discussed the comments and recommendations made at the open forum meetings and answered questions. Anyone who wished to speak was heard.

"We sought to develop a dialogue with the membership at the meetings," Smith said. "We informed the membership that we were still open to suggestions and changes if they found anything that was particularly objectionable." Acting on their own, some states adopted resolutions through their statewide associations expressing support for or disagreement with the committee's work. "We received many such resolutions. Generally, we found overwhelming support for the approaches we were formulating."

Interest in the meetings was high. NRECA reported the largest turnout for regional meetings in its history to date. The combined registration was 5,912, almost 10 percent higher than the 5,421 average of the previous five years. The previous high attendance was 5,633 in 1963.

Most of the comments and discussion were about the financing proposals. There was concern about higher costs, but participants generally favored the self-help approach.

One theme that was frequently sounded was the need for program unity. "Some of us think we can go it alone. We can't," said a participant in the Region 3 meeting in Louisville, Kentucky. "We failed in our legislative effort to obtain supplemental financing largely because we did not have cooperation within the program. If our power supply program is lost due to non-cooperation, the rural electric program will pass from the scene."

A consensus of support for the self-help approach emerged. "We were encouraged by the attendance. We came back with a strong affirmative vote from the people who heard the presentation," Slatt said. "I believe this was the point where we realized that all systems were 'go.' "

Final Decisions

The committee continued to debate its recommendations. It made most of the final decisions at the November 13–15, 1968, meeting in Washington, D. C.

Up to this point, although the committee had moved toward a consensus, it kept open options for various legislative and self-help approaches. Kermit Overby, director of legislation and research for NRECA, reported on the outlook for the 91st Congress. Richard M. Nixon had just been elected president, but there was little change in Congress. Overby said that there was danger in pressing for supplemental financing legislation. "We had a good chance of getting a bad bill vetoed as a last resort under a Johnson administration, while this is an unknown quantity under a Nixon administration," he said. He also noted that the Department of Agriculture and the Budget Bureau (predecessor of the Office of Management and Budget) were not likely to be sympathetic in supporting legislation. On the other hand, there was an opportunity to sell the new administration on new approaches to deal with rural problems, and a more independent role might be possible.

Legal Consultant Charles Samenow presented a report on the laws under which an independent financing institution could be organized and a structure for an institution organized under the cooperative law of the District of Columbia. Katzin reported on the financial aspects of the lending institution and its projected lending capabilities under various scenarios.

On Thursday, November 14, 1968, the committee unanimously agreed to accept the recommendation of the Financing Subcommittee that it proceed with development and documentation of a plan to implement a self-help financing program. Smith asked the subcommittee to present a detailed plan for consideration by the full committee at its December meeting.

The subcommittee had already discussed many of the issues it needed to address. "We started right from the beginning, considering such questions as: Should it be a nonprofit organization? Should it be a corporation for profit? Should it be organized as a municipal, tax-exempt, issuing entity? If it were to be incorporated, where should it be incorporated? Delaware? Kansas, the heartland of America? Or the District of Columbia?" Katzin said.

Katzin gave credit to Samenow for helping to conceptualize what the entity would be. "I don't think that we should pass up the amazing contributions and support that Charlie brought to the picture right from the beginning," he said. "His intimate knowledge of the legal background and tax situation of the cooperatives was most valuable."

The decision to create a nonprofit, cooperative organization incorporated in the District of Columbia came fairly easily. "The members mandated that the organization be member owned and member controlled," Smith said. "It must always be sensitive to the needs and requirements of its member rural electric systems. It should provide capital at the lowest cost possible, and this capital should be supplemental to the existing loan program. It should be cooperative in structure, nonprofit in its operations, and provide maximum flexibility in all its loan programs." Organizations eligible for membership included cooperatives, nonprofit corporations and public bodies eligible to receive loans from REA, along with organizations owned or controlled by such organizations. The new cooperative was to be incorporated in the District of Columbia because of the presence of REA and NRECA and because of a favorable legal climate.

Balanced Member Representation

The board would include representatives from ten districts congruent with NRECA's ten regions and an eleventh district comprised of the District of Columbia. NRECA was the only potential member in District 11. Some cooperatives believed that only managers should serve on the board, whereas others believed that only directors should serve as directors for the new organization. "We solved that one by having two directors —a manager and a director—from each region," Slatt said. "However, no state could have two directors." The form of representation for District 11, or NRECA, continued to be debated until shortly before the final report was presented at the annual meeting in March 1969.

The committee believed it was essential to keep the election process entirely in the hands of the membership, from beginning to end. They devised a procedure where candidates would be nominated through elections at the district level and elected in national elections.

The committee decided to limit the terms of directors to two three-year periods. "We weighed this carefully, because we had to decide whether it would be better to have a director leave after six years and go out into the country as a firm supporter of CFC versus the loss of the experience gained over six years as a director," Slatt said. "Although it wasn't a unanimous decision, it was a very one-sided vote. As I look back at all the decisions we made, I believe that was probably one of the wisest."

One reason for the committee's decision was concern about the consequences of a lack of turnover on the board. "A lot of the electric cooperatives were started in the late 1930s and 1940s," Shesser said. "Here we were in the 1960s, and the same people were sitting on the boards who were sitting there when the co-op was founded. These were all well-meaning, dedicated individuals, but the co-op they were involved with in the early days was no longer the same organization 30 years later. It was a more complicated, more sophisticated operation, and a lot of the directors didn't keep up to date. There was a crying need for turnover. We hoped that CFC would become an example for the rest of the program and that the rest of the program would adopt similar requirements."

From the very beginning, the CFC leadership emphasized training for its directors and encouraged them to be involved thinkers. "We believed that if the board of directors of CFC strictly concerned themselves with policy matters and left the managing to the key staff, it wouldn't take a new director too long to get up to speed," Slatt said.

Other issues had to do with the basic financial structure of the institution. "How would we obtain the capital subscriptions? How would those subscriptions be represented? How would the capital grow? How much leverage could we build into this concept? I was responsible for making sure that we were creating creditworthy instruments that would meet the traditional standards and tests of the marketplace," Katzin said.

Member Equity Investments Key

A key element of the plan called for the systems to invest a substantial sum in equity capital for 50 years. "If the systems agreed to this, other proposals were open to compromise," Smith said.

The cooperatives had an obvious source of funds for the equity investments—more than $346 million in prepayments on debt to REA by 1969. REA policy required the systems to prepay their outstanding obligations if they generated surplus funds. "This money was worth 2 percent to them because they were borrowing money at 2 percent," Katzin said. "If they paid it back, they were getting credit on a 2 percent obligation. If we paid them 3 percent for leaving the money with us rather than the 2 percent they would earn if they gave it back to REA, they would have a good financial incentive, a 50 percent gain, for investing in CFC."

Katzin wanted the investments to be permanent capital, like common stock. "Because of the need to revolve capital credits, there was no support for the idea of making them a permanent investment," Katzin said. Instead, the committee established a term of 50 years. "It seemed far enough in the future at that time to give us some assurance that they would extend beyond the life of any initial loans we would make. We established a term and the members provided the capital. That's how we got the name Capital Term Certificates [CTCs]." Each cooperative would receive a certificate to represent its investment. The CTCs were to be subordinated to all outside financing. The committee established a formula and schedule for collecting the basic CTC investments, called Subscription CTCs, over a 15-year period. Borrowers would also be required to purchase Loan CTCs as a condition of loans.

The Long-Range Study Committee met again on December 10 and 11. One issue that had to be addressed was the new organization's relationship with NRECA and NRECA's representation on the board of directors. Some members of the NRECA staff believed that CFC should be organized as a department of NRECA rather than as a separate organization. "They recognized, and it was a valid concern, that CFC would control the money. That would make it a very powerful organization. They were concerned that it would supplant NRECA," Shesser said.

On the other hand, one of NRECA's major functions was to support its members' position on political issues before Congress and other legislative bodies. "What we needed was a nonpolitical entity," Herriott said. "CFC has always had all the trappings of a conservative financial organization, and that is what we need to raise money for us on the most advantageous terms." The committee was concerned that NRECA would need to take positions on issues about which CFC would prefer to remain silent.

Katzin offered the strongest argument for an independent organization. He said that Wall Street would be hesitant to raise money for a financing organization that engaged in partisan politics.

The solution was for CFC to be independent of, but maintain close liaison with, NRECA. "The institution's bylaws should include language clearly restricting the institution from competing with NRECA services," the committee's resolution said.

"We all agreed that NRECA and CFC were going to have to work closely together," Herriott said. "Also, CFC was going to have to depend on NRECA to accomplish its legislative objectives."

To further strengthen its ties to NRECA, the committee voted to give District 11 five seats on its board.

Naming CFC

By all accounts, the committee spent at least half a day discussing a name for the new organization. "Because it would be national in scope, the word 'national' should be included," Smith said. "We discovered in our discussions with investment banking firms on Wall Street that 'rural' was very appealing terminology. They felt that the rural areas of the country were stable and sound and that rural people paid their bills. We were financing utility systems, so we added the word 'utility.' It was a financing organization so the term 'finance' must be included. And finally, it was going to be a 'cooperative.' " National Rural Utilities Cooperative Finance Corporation it would be.

The name National Rural Utilities Cooperative Finance Corporation was quickly shortened to CFC in everyday use.

Nobody expected to use the long name very much anyway. "We recognized even as we finally settled on a name that it would be shortened to CFC and that the abbreviation would work very well," Bob Partridge said.

The telephone cooperatives were in the process of establishing the Rural Telephone Bank at the time, but the National Telephone Cooperative Association had asked the committee to include provisions that would allow the new organization to make loans to rural telephone cooperatives in the future. "We made the structure we were recommending broad enough to include telephone financing, because at that time there was some question about the structure of their bank," Smith said. "It was understood that CFC could not do this without additional investments in equity by the telephone cooperatives and without authorization from the membership. Even though telephone financing was not included in the initial stage, the door was left open so it could be included if necessary."

On January 7, 1969, the group met to polish their final report and deliberate last minute details. The discussion emphasized two points that must be conveyed to the membership: the new organization was in no way a substitute for the REA program, and NRECA and CFC needed to work harmoniously in the future. After a little fine-tuning, the committee was ready to present its final report.

On January 8, 1969, the NRECA board of directors approved the report and directed that it be presented to the NRECA membership for action at the annual meeting in Atlantic City, New Jersey, on March 16–20, 1969.

The committee immediately sent a copy of its report and

recommendations to each rural electric system and conducted a cross-country tour the week of January 27 to hear the final comments and suggestions of the membership.

There was still some opposition to the supplemental financing concept. Slatt recalled that at one of the fly-in sessions, an individual who didn't think supplemental financing was a good idea urged all those who opposed it to get up and follow him out of the meeting. "Fortunately for us, the only one who

Delegates to NRECA's annual meeting approved the formation of CFC on March 17, 1969.

got up and followed him was his own attorney," he said.

Two issues drew a significant amount of member comment and suggestion. One was a proposed 3 percent premium for late membership in CFC. The systems said that amount was too low and suggested raising it to 4.5 percent.

The other issue was representation of NRECA on the CFC board. The membership expressed concern that NRECA, as the only member in District 11, would have five directors, whereas the other districts would only have two, and that the

District 11 directors would not be subject to the same age and tenure restrictions as other districts. The committee discussed alternatives but deferred action until the next meeting.

The final sessions of the Long-Range Study Committee were held during the NRECA annual meeting in March 1969. On March 14, the committee again discussed at length the two issues raised by the membership at the January fly-in meetings: the penalty for late membership and the District 11 board representatives. It also met with the NRECA board to solicit its views. Based on the discussion with the NRECA board, the committee then adopted a motion to revise its plan to give two directors to District 11 and to provide for a premium of 4.5 percent, compounded annually, for late subscribers to CFC.

The committee met briefly for one last time on March 17 before the presentation of its Final Report and Recommendations to the NRECA membership that afternoon.

"This has been the most thorough and dynamic effort I have witnessed within the rural electric program in the past 30 years," Clyde Ellis told the committee.

"I will be working to see that North Carolina is the first state to have 100 percent of its rural electric co-ops become members of the National Rural Utilities Cooperative Finance Corporation," said committee member Cecil Viverette.

"Oklahoma accepts the challenge from North Carolina," responded his colleague Lloyd Patton.

Then the committee adjourned for the last time.

A few hours later the membership of NRECA approved the organization of CFC with fewer than 20 dissenting votes. On October 1, 1970, North Carolina and Oklahoma became the first two states with more than two cooperatives to achieve 100 percent membership in CFC.

PART 2 CFC GETS STARTED

The first three years of CFC's existence were devoted to establishing a strong base for its future operations. During this time, CFC acquired a staff, a membership and an equity base. It structured the documents and procedures needed for a viable working relationship with REA. It outlined the policies and procedures for its loan programs and made its first loans. It developed sound relationships with the financial community, achieved an A credit rating and began its financing programs. By the time CFC issued its first collateral trust bonds on December 7, 1972, it had everything it needed to obtain funds from the private capital markets and lend them to its members at a reasonable cost.

Highlights

CFC was incorporated on April 10, 1969.

A total of 512 systems qualified as founding members of CFC by paying their membership fee before September 1, 1969.

The Internal Revenue Service granted CFC's request for tax-exempt status as a social welfare organization under Section 501(c)(4) of the federal tax code on October 30, 1969.

CFC's first annual meeting took place on March 20, 1970.

On December 8, 1970, CFC approved its first short-term loan for $500,000 to Warren County Rural Electric Membership Corporation in Bowling Green, Kentucky, at an initial variable interest rate of 7 percent.

REA issued Bulletin 20-17 setting forth the conditions under which rural electric cooperatives would be required to obtain supplemental financing on February 1, 1971.

CFC made its first two long-term concurrent loans at a fixed interest rate of 7.5 percent for 35 years during the annual meeting in February 1971: $100,000 to North Arkansas Electric Cooperative in Salem, Arkansas, and $39,000 to Washington County Electric Cooperative in Sandersville, Georgia.

CFC issued its first collateral trust bonds, Series A, a 35-year $50 million issue priced at 7.4 percent, on December 7, 1972.

After the membership's approval of the creation of the National Rural Utilities Cooperative Finance Corporation during its 1969 annual meeting, the NRECA board appointed a group of men to serve as incorporators of the new organization and act as its board of directors until elections could be held. Most of the CFC board members had also served on the Long-Range Study Committee, so there was a strong sense of continuity and purpose. The group held its first meeting in Atlantic City on March 18, 1969, but due to a legal technicality, the actual incorporation of CFC did not take place until April 10 in Washington, D. C.

CFC's first board of directors, its consultants and staff worked hard to be sure that the new organization succeeded. Pictured on November 10, 1970. **Back row, left to right:** ***Paul Fox, Milton A. Chase, Ira Shesser, Robert D. Partridge, Charles U. Samenow, Louis B. Strong, E. V. (Brick) Lewis, Arden Warner, Raymond W. Rusteberg, Jerome Katzin, James T. Dudley, Walter Harrison, Leland Kitt, Owen Manning, Earl L. King, Roger Oresman and W. C. Carlton.*** **Front row, left to right:** ***John Pierce, J. R. Cobb, Sailey Ennis, Lloyd Patton, Joe Meng, Vincent P. Slatt, J. K. Smith, Clarence Peterson, Robert T. Beall, Virgil Herriott and Leland Leatherman.***

At that point, CFC had a mission but none of the things needed to carry it out. There was no organizational structure—no staff, no office, no bylaws, no money and no membership. The new CFC board acted quickly to get itself organized. J. K. Smith was elected president with Vincent P. Slatt vice president and Leland Leatherman secretary-treasurer. The board adopted a budget of $126,000 for the April 1– December 31, 1969, period.

Charles U. Samenow agreed to serve as consulting corporate counsel. Harry Mitchell, NRECA's controller, was appointed CFC assistant treasurer and Ira Shesser assistant secretary. Mitchell managed the money during the interim period while Shesser provided the staff and administrative assistance.

The board also established relationships with two organizations that would be very important to CFC's future. Jerome Katzin was retained as financial advisor, and his firm, Kuhn, Loeb & Company, was designated managing underwriter. Katzin recommended Roger Oresman as legal counsel, and the board appointed his firm, Milbank, Tweed, Hadley and McCloy, as general counsel.

Many members of the Long-Range Study Committee served on CFC's first board of directors and as incorporators of CFC.

William A. Byers *
Director, Grand Valley Rural Power Lines Grand Junction, Colorado

W. C. Carlton
Manager, Carteret-Craven Electric Membership Corporation Morehead City, North Carolina

J. R. Cobb
Manager, Texas Electric Cooperatives Austin, Texas

James T. Dudley
Manager, Singing River Electric Power Association Lucedale, Mississippi

Sailey Ennis
Manager, Washington Electric Cooperative, Inc. East Montpelier, Vermont

Walter Harrison
Manager, Georgia Electric Membership Corporation Millen, Georgia

Virgil Herriott
Manager, Sioux Valley Empire Electric Association Colman, South Dakota

T. W. Hunter+
President, National Rural Electric Cooperative Association
Attorney, Newberry Electric Cooperative Newberry, South Carolina

Earl L. King
Manager, Iowa Association of Electric Cooperatives Des Moines, Iowa

Leland Kitt
Treasurer, Southwest Public Power District Palisade, Nebraska

Leland F. Leatherman
Attorney, Arkansas Electric Cooperatives, Inc. Little Rock, Arkansas

E. V. Lewis
Manager, Central Electric Power Cooperative, Inc. Cayce, South Carolina

Owen Manning
Manager, The Frontier Power Company Coshocton, Ohio

R. B. Moore†
Manager, Lea County Electric Cooperative, Inc. Lovington, New Mexico

Robert D. Partridge *
General Manager, National Rural Electric Cooperative Association Washington, D. C.

Lloyd C. Patton †
Director, Cotton Electric Cooperative Chattanooga, Oklahoma

Clarence W. Peterson
Manager, Lake Region Co-op Electrical Association Pelican Rapids, Minnesota

John A. Pierce *
Director, Raft River Rural Electric Cooperative, Inc. Malta, Idaho

Raymond W. Rusteberg *
President, Association of Illinois Electric Cooperatives
President, Monroe County Electric Cooperative Waterloo, Illinois

Vincent P. Slatt
Manager, Inland Power and Light Company Spokane, Washington

J. K. Smith §
Manager, Kentucky Rural Electric Cooperative Corporation Louisville, Kentucky

Arden Warner *
President, Kosciusko County Rural Electric Membership Corporation Warsaw, Indiana

** Did not serve on the Long-Range Study Committee.*

+ Replaced by Louis B. Strong in 1970.

† Not a CFC incorporator.

§ Replaced by Joe Meng when Smith became governor

"I wanted Roger in the beginning, because his firm was general counsel for the Chase Bank," Katzin said. "I was very concerned that we get some commercial banking backup." Milbank's relationship with Chase gave CFC credibility in the banking community. Through his relationship with Philip Sporn and the Buckeye project, Katzin was also able to introduce CFC and the rural electric program to Manufacturer's Hanover Trust Company, American Electric Power Company's lead bank.

Three men who provided valuable staff assistance to the Long-Range Study Committee agreed to assist the new CFC board of directors in organizing the corporation. Pictured,* left to right: *Ira Shesser, assistant secretary; Harry Mitchell, assistant treasurer; and Charles U. Samenow, consulting corporate counsel.

In addition, the board appointed several committees to address its most immediate concerns: the Operations Committee, chaired by J. R. Cobb; the Finance Committee, chaired by Virgil Herriott; the Membership Committee, chaired by James T. Dudley; the REA Coordinating Committee, chaired by Vincent P. Slatt; the Bylaws Committee, chaired by Leland Leatherman; and the Personnel Committee, chaired by J. K. Smith.

Board Establishes Target Dates

The board approved the bylaws, arranged for financing and designated individuals who could incur expenses and sign checks from the new checking account, approved a corporate seal, adopted a fiscal year, applied for membership in NRECA, and put together a framework for keeping the potential membership informed of its activities. It also established a series of target dates for 1970:

- July 1: Obtain REA concurrence with membership procedures.
- July 1 to October 1: Conduct membership drive.
- October 1: Determine adequacy of rural electric system participation.
- November 1: Employ chief executive officer.
- December 1: Open office.
- December 1: Collect first CTC investments.

It was an ambitious schedule, but the board met or improved on each target.

Without the support of NRECA, these achievements probably wouldn't have been possible. During the five years before CFC's incorporation, NRECA devoted a significant portion of its staff time to the development of a supplemental financing program and spent an additional $143,000 on other resources. It agreed to provide temporary office space, staff and related management services for an interim period until January 1, 1970, charging only for direct costs such as postage and copying. The NRECA board of directors authorized a $90,000 loan to CFC to meet those and other expenses.

CFC had three important tasks to complete before it could begin its business operations:

- It needed to obtain tax-exempt status for federal income tax purposes.
- It needed to negotiate loan policies and procedures with REA.
- It needed to address securities issues regarding the CTCs.

CFC could not solicit investments in CTCs until these issues were resolved.

In the interim, the board needed to measure the level of interest in the new organization. On July 1, 1969, it invited all rural electric systems to submit their applications for membership in CFC by paying the membership fee ($1,000 for distribution and G&T systems, $200 for cooperative associations) with the understanding that this was not a final commitment. The systems would not be asked to make that decision until a sufficient number of systems indicated an interest in joining and the basic issues were resolved. Then CFC would invite the systems to subscribe to CTCs. If a system decided not to participate, the membership fee would be refunded, less a proportionate share of operating expenses for the interim period. This approach gave the systems a way to demonstrate their commitment to CFC while the board and staff addressed the basic issues. For legal purposes, the incorporators served as charter members until there was a rural electric member from each of CFC's 11 membership districts.

The results were encouraging: 512 systems qualified for special recognition as founding members by paying the membership fee before September 1, 1969. "We believed that if 500 systems would agree to execute the subscription agreement and pay the $1,000 membership fee, we would have a viable organization. We could proceed with the establishment of the supplemental financing organization," Smith said. In just two months, CFC had commitments exceeding the minimum requirement.

CFC Seeks Tax-Exempt Status

Tax-exempt status was necessary to the way CFC planned to operate. "That was essential, because CFC had to be able to raise money and lend it to the member cooperatives at the most favorable rates that it could," Oresman said. "If there had to be an income tax levied on its margin, that would make the money more expensive for CFC members."

Although it was possible to operate as a taxable cooperative and avoid most federal income tax liabilities by distributing most of the organization's margins on an annual basis, substantially all of CFC's potential members were tax-exempt at the time. There was an assumption that CFC, too, should be tax-exempt. "If CFC weren't tax-exempt, it wouldn't have been as successful, because it couldn't have raised money as efficiently," Oresman said.

Samenow's initial meetings with the Exempt Organizations

Membership in CFC has continued to grow.

Year	Class A	Class B	Class C	Class D	Associate	Total
1970	715	36	31	1		783
1971	760	31	34	1		826
1972	775	33	33	1		842
1973	784	34	33	1		852
1974	786	37	36	1		860
1975	790	40	38	1		869
1976	792	43	42	1		878
1977	795	45	41	1		882
1978	799	50	46	1		896
1979	802	50	50	1		903
1980	801	57	54	1		913
1981	801	61	55	1		918
1982	802	61	56	1		920
1983	806	60	58	1	1	926
1984	807	60	66	1	2	936
1985	815	62	70	1	5	953
1986	822	63	71	1	7	964
1987	821	65	73	1	8	968
1988	824	65	73	1	25	988
1989	829	65	73	1	40	1,008
1990	829	64	74	1	40	1,008
1991	830	64	74	1	53	1,022
1992	824	62	69	1	59	1,015
1993	830	64	71	1	63	1,029
1994	833	66	70	1	69	1,039

Class A, rural electric distribution systems, Class B, generation and transmission cooperatives; Class C, statewide and regional associations; Class D, national association; Associate, nonprofit corporations owned, operated or controlled by Class A, B or C members or by CFC.

Branch of the Internal Revenue Service (IRS) were discouraging. "The conference yielded no encouragement on the technical aspect," he reported. "However, the importance of exemption to the success of the enterprise was recognized, and it was suggested that application for exemption be filed so that not only the technical side but the policy implications could be examined."

Milbank structured an argument that CFC was entitled to the exemption because it was a social welfare organization. "For some 30 years, the REA was a government program to bring central service electricity to America's rural areas. That was recognized as an important social policy," Oresman said. "When it became clear that the government was not going to be able to supply all the capital that these cooperatives needed in the late 1960s and CFC was created to fill that void, the argument was that this was a supplement to the social welfare purpose."

The argument was persuasive. The IRS granted CFC's request for tax-exempt status as a social welfare organization under Section 501(c)(4) of the federal tax code on October 30, 1969. "If the IRS had not given us this ruling, CFC could not have survived," Smith said. CFC continued to enjoy the exemption as its financing programs and lending activities expanded.

CFC was also able to obtain exemption from the District of Columbia income tax, based on the federal IRS exemption. In addition, it had to obtain an exemption from the District of Columbia usury statute and loan shark law, which limited the interest a lender could charge. This required Congressional action but was not a major hurdle.

Exemptions from Securities Act of 1933 and Securities Act of 1940

It took over a year for CFC to obtain exemption of the CTCs as securities under the Securities Act of 1933. CFC argued that the certificates, because of their 50-year term and relatively low interest rate, were not being sold to people with the expectation of return or profit and, therefore, were not securities within the meaning of the relevant section of the Act. In addition, the certificates were more in the nature of a private placement rather than a public issue, because the organizations that would be investing in the CTCs were members that had been intimately involved in creating CFC. Finally, it was submitted that a section of the Act, which is no longer in effect, exempted the securities of certain types of institutions that confined their business to making loans to members.

The SEC granted a no-action letter exempting the certificates from the registration requirements of the Securities Act of 1933 on April 17, 1970. "It meant that we didn't have to prepare a registration statement and the prospectus that would go with it, which would have had to be cleared by the SEC," Oresman said. "It would have been a somewhat longer process that we didn't want to have to go through. The exemption was easier, and it was less expensive. It also helped with the states in the blue sky area."

Then CFC had to obtain an exemption from the Investment Companies Act of 1940, which regulates mutual funds and other investment companies. "Technically, a company that holds the kinds of mortgage notes of the members that we were going to have would be an investment company that gets regulated under very stringent laws with many restrictions on what it can do," Oresman said. CFC wouldn't be able to operate as it wanted to under that type of regulation.

Unfortunately, the SEC office that determined the exemptions under the 1933 Act did not determine the exemption under the 1940 Act. "We had to go over a lot of the same material," Oresman said.

The process was time-consuming for two reasons: one, it is the nature of regulation, and two, many of the issues were questions of first impression. That is, CFC was a new type of organization. The regulators had not decided a similar case in the past. "They have their own chain of command for things that are novel, and so it took some time. Also, they had many questions," Oresman said.

The SEC issued a no-action letter effectively exempting CFC from the Investment Companies Act of 1940 on July 7, 1970. Without these exemptions the initial financing activities of CFC, including the solicitation of systems to subscribe to CTCs, would have been severely limited and even more complex.

CFC also had to resolve various state legal issues, including "blue sky laws" regulating the sale and issuance of securities, before it could ask members to invest in CTCs. In some states, the membership had to approve a cooperative's membership in CFC, and the bylaws of many systems had to be modified so that CFC could make loans to them. This could usually be accomplished at an annual meeting.

Nineteen states required membership approval by a vote of two-thirds or half of all members for an electric cooperative to borrow from secured sources other than REA. Five states had statutes that limited the term and interest rate on loans to cooperatives. All of this had to be changed. Fortunately, CFC received excellent support and cooperation from the cooperatives and statewide organizations involved as well as from their state legislators in doing the things necessary for CFC to do business in their states, but it took many months and much effort.

Hamil Backs CFC Plan in Word and Deed

CFC's negotiations with REA were even more lengthy. David A. Hamil returned as administrator of REA in March 1969 and gave his full support to the development of CFC. "This isn't the plan that I had in mind a few years back, but it is a plan," he had told the Long-Range Study Committee. "If this is the one that the people want to follow, I will use every ounce of energy that I have and the influence of the Rural Electrification Administration as well to make it work." Hamil kept that promise, but there were many issues to work out between REA and CFC.

CFC had to obtain the approval of REA for its borrowers to invest in CTCs. CFC and REA had to agree on a common mortgage, probably the most difficult issue. Finally, the REA administrator had to designate which systems would be required to obtain supplemental financing. Each of these broad issues encompassed numerous details that had to be addressed.

"There were little differences of opinion as we went along," Hamil said. "It was recognized that J. K. Smith and I had an understanding that if our people got into a log jam, they weren't to engage in a fist fight, but they were to let us know. J. K. and I would then get in the room without them, and independently we would arrive at a position. Then we would say 'this is the way it's going to be.' And I think it's safe to say that that's the way it was." Without the support of Administrator Hamil and the close working relationship that developed between REA and CFC staff, these issues would not have been resolved.

Although Hamil set the tone, it was the REA staff who had to find a way to make things work. "Ed Wilson, my boss when I was at REA, was the one who spearheaded it through," said

Charles B. Gill, later CFC governor. "He gave direction to most of the staff. Lou Gorrin was also involved early on with the mortgages and so forth."

Administrator Hamil met with the CFC board during its second meeting June 30 to July 1, 1969. "I am confident that REA can work out satisfactory arrangements with CFC," he said. He reported that he had met with members of Congress to tell them about CFC and its role in providing supplemental financing for rural electrification. He said that he had also consulted with Nixon administration officials, who had raised no difficulties.

After the board meeting, he publicly announced that REA favored the establishment and operation of CFC. He sent information on CFC to all of REA's field representatives and key personnel. "REA has endorsed CFC," he told them. "CFC right now is our best hope to bring urgently needed capital into our electric program."

"Dave Hamil was very instrumental in developing some of the solutions to these problems," Smith recalled. "He informed the REA staff and the general counsel's office that he would like to see this matter worked out on a fair and equitable basis so CFC could become operational."

However, there were early questions about REA's authority to do the things CFC wanted and needed it to do. Katzin was very surprised when Administrator Hamil said he needed to take some of the issues back to Congress. "We were well aware of the unhappy experience we had had previously. We felt that anything we were seeking from REA could be accomplished under the administrator's rule-making authority," Katzin said. "Nevertheless, the administrator felt that it was probably better politics for the program that Congress have prior knowledge of what we were doing. As a result, we made a number of presentations before the Congressional Appropriations Committee." No specific legislation was ever required, and the Appropriations Committee was generally supportive.

CFC engaged former REA employee Robert T. Beall as a consultant to help coordinate CFC's policies and procedures with REA. Beall, who had been one of REA's first employees in 1935, had structured REA's loan system and knew REA as well as anyone. He helped CFC find ways to integrate its program quickly and easily.

REA could not formally approve a borrower's investment

in CTCs until a system requested it, but Hamil signaled REA's willingness to approve CTC investments in a new general funds bulletin, 1-7, issued in the fall of 1969. The policy limited general funds to the greater of 8 percent of total utility plant or $100,000, but Section IV of the bulletin allowed exclusion of certain investments. "Investments in, and commitments to invest in supplemental financing arrangements, including CFC, within a 12-month period can be excluded," the transmittal letter said. In other words, systems could invest in CFC without affecting their ability to accrue general funds for other purposes.

Additional support for investment in CFC's certificates came from Congress. The report of the House and Senate Appropriations Committees for fiscal year 1971 included a recommendation that principal repayments on REA loans be deferred beyond the then-current three-year allowance with the understanding that borrowers requesting and receiving extended deferral would invest an equal amount in CTCs. In other words, systems could delay payments to REA in order to invest in CTCs. Each district endorsed this recommendation at the fall 1970 regional meetings.

Common Mortgage Defined REA-CFC Relationship

Resolving the CTC issue was relatively easy. Negotiating the REA-CFC common mortgage was not, and it was the most important element to CFC's operation. "The mortgage was the point of greatest impact, because our interest in the mortgage was the security that we were going to be offering the investing public," Oresman said. "We had to reach consensus on the form of mortgage that the investment bankers were satisfied would provide the kind of security that would enable them to sell CFC collateral trust bonds."

The mortgage became the cornerstone of the relationship between REA and CFC, and each organization had to be sure it contained the elements it needed to function. "The mortgage has been the subject of continuing negotiations, and even momentary conflicts, between REA and CFC," former CFC Corporate Counsel Louis A. Roland said. Before joining CFC, Roland held the position of deputy director of REA's Office of General Counsel. He worked on both sides of the mortgage issue. "REA has had to preserve certain governmental interests and rules that limited REA's discretion. On the other hand, CFC had its own interest on behalf of its members, and these interests had to be protected in order that CFC achieve its goal of becoming a viable and useful financial organization."

Hamil recognized that REA was going to have to accommodate, or share, its lien on cooperative systems to CFC in order for the new organization to raise funds. That was something REA had never done. "CFC was going to have an opportunity to develop so it could meet its financial responsibility. That was the judgment at the time, and I have no qualms about what we did," he said. "We had to give concessions to make sure that CFC would be able to fill its assigned role."

The government held a sole, exclusive first mortgage on all the property of the borrower cooperatives, including present and future properties. "REA had, of course, a mortgage that was outstanding, and they didn't want to make a lot of changes in it," Oresman said. "We were concerned about aspects of control, what REA could do without consulting us."

The mortgage provisions, developed over a period of 35 years, were contained in a document almost 40 pages long. It gave the government various controls and remedies in the case of default. "Naturally, the government was reluctant to give up any of these controls and powers that had been worked out over a long period and that seemed to be working pretty well," Roland said. "The organization and procedures of REA were based on its controls and powers under the mortgage."

Early in the negotiations, Oresman suggested they adapt a standard commercial form of mortgage that utilities used. "Of course, REA would not stand for that," Roland said. "The whole REA program had been based on the detailed form of provisions of the REA mortgage, and so the form of existing exclusive REA mortgage became the basis of the common mortgage."

In accommodating its mortgage to CFC or any other supplemental lender, REA's primary concern was to protect the security position of the government. "There were a lot of things that REA has a voice in that you wouldn't find in an ordinary mortgage," Oresman said. "We weren't trying to make the mortgage harsher or stricter but only to allow CFC to have some element of control."

CFC Requires First Mortgage

CFC's position was that it had to become a viable financing organization with the ability to raise money at reasonable rates in the private capital markets, primarily through collateral trust bonds.

For that, it needed security. "Our financial advisors told us that in order to have a marketable security, CFC would have to have a common mortgage that would provide our lien with equal status to that of REA," Smith said. "In other words, a first mortgage was essential."

REA suggested that CFC make unsecured loans or second mortgage loans, subordinating to REA, but that would not have given CFC adequate strength to issue collateral trust bonds. "CFC had to have a mortgage," Roland said. "Unsecured obligations from the borrowers would not be sufficient to enable CFC to get the money or the bond ratings it needed. There was a clash of positions between REA and CFC on this point." CFC also needed to be able to make an independent judgment in the exercise of its powers and its use of remedies under the mortgage and to act alone when necessary.

"When I was with REA, I realized that we obviously had to find answers for the questions that arose from this," Roland said.

First, REA had to find a legal justification for the government to give up some of its rights under its preexisting, exclusive mortgage. Roland researched the General Accounting Office precedents, some of which went back 150 years, for permitting or not permitting a government agency to give up legal rights that the government had obtained. There were numerous possibilities, but REA finally based its mortgage accommodation on a ruling that said that governmental rights could be surrendered, provided that the administrator or the head of an agency found that it resulted in a benefit to the government. "It is an administrative determination," Roland said. "The administrator has to find that the government would benefit from his action, and this finding needs to be supported on a reasonable basis. Only if the finding were completely arbitrary or an abuse of discretion would it be reversed. This was the basis on which the common mortgage was developed."

The REA administrator reasoned that, because it was becoming increasingly difficult for the government to obtain funds to support the rural electric program, it would benefit the government for the systems to obtain private funds. That would relieve the government of growing expenditures, a benefit that justified surrendering certain rights to the party who was lending the private funds.

The initial common mortgage, now called a Type I mortgage,

gave CFC equal rights and remedies with the government across the board. CFC had to act with REA to permit certain things and to waive certain provisions. CFC could exercise equal rights with the government in applying remedies, including the ultimate remedy of foreclosure. The only qualification was a 30-day waiting period before CFC could exercise remedies to permit the government to decide what it wanted to do first.

"To justify this, REA insisted that CFC obtain proof from the financial experts that this first mortgage with these remedies for CFC was necessary," Roland said.

Katzin supported CFC's position in an August 24, 1970, letter to REA. "Under existing legal investment laws regulating investments by institutions in many states, the debentures which CFC proposes to issue would only be legal investments if they were regarded as a 'prudent investment' or as being 'adequately secured,'" he wrote. "Even if debentures collateralized by first mortgage loans would meet such tests, it would be more difficult for the investment committee of an institution to reach such a conclusion if the collateral were a second mortgage. Furthermore, in certain states the legal investment laws specifically refer to 'first mortgage bonds' as the only permitted investments in public utility companies, and investment committees operating under the 'prudent investment' rule might take this as a guide in considering a loan to [investment in] CFC."

That gave Hamil the grounds he needed to proceed. "Once that was established, in view of Administrator Hamil's insistence that we go ahead with the program, we worked out the common mortgage on that basis," Roland said. "This was an acceptable procedure, and it did work. This led to excellent ratings on CFC bonds and provided the strong security needed to support the CFC bonds. Without the common mortgage, I don't think there would ever have been a viable CFC."

In subsequent years, REA and CFC reopened mortgage negotiations to accommodate new financing programs. "As each new program requiring security other than that required for normal collateral trust bond purposes was evolved by CFC, the same problems arose between REA and CFC," Roland said.

As the member systems became interested in more sophisticated financing alternatives and CFC found ways to provide them, the mortgage negotiations with REA became more difficult. "We were always looking for a shared first mortgage, and REA began backing away from their willingness to always give us a first mortgage lien accommodation," Ira Shesser said. REA offered various second mortgage positions for new financing programs. "Some of REA's more conservative staff were beginning to express concern that they were weakening the government's security position. They believed that it was their responsibility as government employees that under no circumstances would the government's loans in these matters be put at risk."

CFC and REA negotiated different mortgage security arrangements to secure different loan and guarantee programs.

Type I

CFC shares a first lien with REA on all assets and revenues of the borrower. CFC, as mortgagor, must approve certain actions of borrower. Exercise of remedies fully available 30 days after a default.

Type II Standard

CFC shares a first lien with REA on all assets and revenues of borrower. Controls exercised by government. Exercise of remedies available 30 days after a default if CFC holds at least 25 percent of the notes.

Type II A

CFC shares a first lien with REA on all assets and revenues of borrower. Controls exercised by government. Exercise of remedies available after 90 days in the event of bankruptcy or nonpayment. The Type II A mortgage replaced the Type II Standard mortgage in 1983.

Type II Pollution Control

CFC shares a first lien with REA on all assets and revenues of borrower. Controls exercised by the government. Exercise of remedies available to CFC 360 days after a payment default to enable CFC to take possession and operate the system; exercise of all remedies available 720 days after a default.

Subordinated Mortgage

Mortgage on all assets and revenues of borrower junior to REA mortgage and any person in whose favor REA subordinates its mortgage. Thirty days after a payment default, CFC can declare outstanding principal and accrued interest to be due and payable immediately. For other defaults, such action can only be taken if REA has taken similar action under the REA mortgage.

Type III

CFC has a first mortgage lien on specific property only. No controls on actions of borrower. Exercise of remedies fully available with respect to the specific property.

In many cases, CFC was forced to accept a lesser mortgage position if it wanted to offer a new program. "We decided we had the financial resources to make the loans. CFC was there to help its member, and if we didn't yield, the party that suffered was the member," Shesser said. For that reason, CFC agreed to accept different mortgage security arrangements to secure different loan and guarantee programs.

Loan Ratios Established

The final element of the negotiations with REA was determining which systems would be required to obtain loans from a supplemental lender and for how much. Nothing in the Rural Electrification Act said that a borrower had to obtain a portion of its funds from a supplemental lender. This was established by an REA policy requiring eligible systems to participate in supplemental financing and stating that, if a system met certain financial tests, then a percentage of its loans must come from a supplemental source. The policy also provided that once a system became a supplemental borrower, it would always be a supplemental borrower. REA hoped to prevent systems from managing their affairs in such a way as to remain or return to being a 100 percent REA borrower.

"This was a courageous move on the part of REA, because it made it clear that from that point on, the eligible rural electric systems were going to be supplemental borrowers," Smith said.

Katzin wanted CFC to be the sole supplemental lender. The creators of CFC believed that the rural electric cooperatives could best meet their financing needs by supporting each other. "We felt that it was important that there not be competing institutions," he said. There was a concern that other institutions would "cherry pick," or provide loans only to the stronger systems. If the stronger systems did not support CFC, that, in turn, would make it more difficult for CFC to meet its members' needs.

Hamil disagreed. He said that the role of government was not to discriminate but to make general rules. As a result, REA did not require that the supplemental source be CFC. "The administrator could not make CFC a chosen instrument," Katzin said. "He [Hamil] recognized that any arrangements he made with CFC could not be tailored exclusively to CFC but would have to be of general application and be available to any other entity that came forward and that could meet the basic requirements established by any actions he took." In retrospect, the issue had little impact, other than lengthening the time needed to develop documentation. "That need of theirs, which we could understand, slowed REA and made them insist on things we really did not feel were necessary," Katzin said.

The original mortgage did give CFC a right of first refusal for additional loans, a right that was never exercised. "We lost some members because of that," said Gill. "If we had recognized early on that you can't lock people in, we would have been better off."

Administrator Hamil saw an advantage if other lenders wanted to provide credit for REA borrowers. "My thinking was that the more sources of capital that we could get involved in the program, the better off our systems would be financially," he said.

Vincent P. Slatt, manager of Inland Power and Light Company in Spokane, Washington, served as Financing Committee chairman on the Long-Range Study Committee and became president of CFC after J. K. Smith was appointed governor.

Establishing Loan Ratios

Finding a way to allocate supplemental funds based on the ability of the systems to pay was a difficult task. "This was really a very serious problem, because we were talking about what people would have to pay for borrowed capital," said Vincent Slatt, who served as chairman of the REA Coordinating Committee. "I don't know how many formulas we considered, but there was always some objection to each one. It began to look like we had come to a real stumbling block."

Finally, Slatt came up with a formula for the plant revenue ratio, or PRR, that essentially deducted the power cost from the cooperative's revenue, then divided it by the total plant. He applied the formula to the cooperatives in his home state of Washington. "I knew the financial strength of each one, who could pay a little bit more and who would have to have 100 percent REA financing," he said. "For example, one cooperative had been operating in the red for 26 years in succession. You couldn't ask them to pay very much more for their borrowed capital. Poor management hadn't caused this system's financial difficulties. The system was divided into two separate parts located 60 to 70 miles apart. This presented a terrible operating problem." According to Slatt's PRR, they qualified for 100 percent REA financing.

Committee member Louis B. Strong, manager of Kay Electric Cooperative in Blackwell, Oklahoma, and president of NRECA, applied the formula to Oklahoma and also thought it would work.

There was only one problem. "REA didn't accept it," Slatt said.

REA's objection was that the formula produced an almost infinite number of graduated steps. If the formula calculated a PRR of 9.4, a system would borrow 94 percent of its funds from REA and 6 percent from CFC. If it came out 4.5, it would borrow 45 percent of its funds from REA and 55 percent from CFC. This, REA said, was too complicated. There was also concern that the calculation was not sufficiently precise to justify such small incremental changes. REA wanted to treat all supplemental borrowers alike.

$$PRR = \frac{\text{Total Utility Plant}}{\text{Operating Revenue and Patronage Capital} - \text{Cost of Power}}$$

PRR	REA Percentage of Total Loan Amount	CFC Percentage of Total Loan Amount
9.01 and above	90	10
8.01 to 9.0	80	20
8.00 and below	70	30

The plant revenue ratio (PRR) was used to determine which distribution systems would be required to obtain supplemental financing and what ratio of REA-CFC funding they would borrow.

The eventual compromise was to adopt the PRR formula but to limit the gradations of supplemental borrowing. Some systems were not required to borrow any supplemental funds. For the rest, the initial participation ratio allowed systems to borrow 90 percent of their funds from REA with 10 percent coming from a supplemental lender. A three-tiered approach eventually evolved, based on 90–10, 80–20 and 70–30 ratios.

The PRR was a fair and equitable way to determine which systems would be required to participate in supplemental financing, but there were those who still opposed any form of non-REA financing. "Some of the systems that were eligible to borrow all their money from REA decided they would not join CFC, because it might jeopardize their ability to borrow from REA," Smith said.

The Invitation to Subscribe

With the basic legal, securities and REA issues resolved, CFC was ready to move to the next stage. On July 1, 1970, CFC issued a formal invitation to the rural electric distribution systems (applicants for Class A membership) and generation and transmission cooperatives (applicants for Class B membership) to subscribe to CTCs. At that time, 987 systems were eligible for Class A and B membership, and 751 had paid the initial application fee. Only

Member investments in Capital Term Certificates (CTCs) provided the equity base CFC needed to raise debt from the private capital markets.

Year	Subscription CTCs (in millions)	Loan CTCs (in millions)	Total (in millions)
1971	$118.1		$118.10
1972	122.5	$0.9	123.40
1973	141.6	3.6	145.20
1974	161.0	8.4	169.40
1975	182.5	17.9	200.40
1976	209.5	22.5	232.00
1977	240.3	27.0	267.30
1978	276.3	33.8	310.10
1979	316.6	47.5	364.10
1980	363.2	62.1	425.30
1981	416.1	76.5	492.60
1982	476.0	87.8	563.80
1983	548.9	94.0	642.90
1984	628.6	111.4	740.00
1985	628.7	123.0	751.70
1986	635.2	133.4	768.60
1987	635.2	153.0	788.20
1988	636.0	192.1	828.10
1989	636.7	216.8	853.50
1990	637.0	233.1	870.10
1991	637.2	255.6	892.80
1992	637.4	273.2	910.60
1993	640.5	292.7	933.20
1994	640.5	311.5	952.00

by agreeing to subscribe to CTCs could those systems become members of CFC and participate in its programs. In addition to investing in Subscription CTCs, the member systems would also be required to invest in Loan CTCs as a condition of loans. CFC maintained separate classes of membership for statewide and regional associations (Class C) and NRECA (Class D), which were not required to invest in CTCs. The initial CTC calculations were based on operations in the year 1967, before the concept of CFC was developed, to ensure fairness and the inability of any system to manipulate CTC obligations.

The invitation to subscribe to CTCs presented a complete plan for CFC's operations to potential member systems. In addition to describing the CTCs, incentives for early payment and penalties for late payment, it outlined basic CFC operations and lending policies. The cutoff date for subscribing without penalty was October 1, 1970, but the board agreed to consider exceptions to that and other requirements that might result in a hardship to a particular potential member. At the time, CFC projected receiving $350 million in CTC subscriptions over a 15-year period.

The board scheduled a series of fly-in meetings for the week of July 27 to answer questions and generate support. The full board and its consultants participated. More than 2,300 rural electric leaders attended. The consensus of the board was that its information was well received. It was now up to each system to make a decision.

Staffing Up

While these activities were underway, the board turned its attention to hiring permanent staff. The first priority was to select a governor, the title carried by CFC's chief executive officer. "The title was my suggestion," Katzin said. "For image purposes, I wanted to liken the chief executive officer to the governor of the Federal Reserve System, which was, and still is, the highest level in financial community functioning. It was kind of a unique term, and it took awhile for people to get used to it." The head of the FCS also carried the title governor.

In November 1969, the board engaged Nelson, Nichol & Company, Inc., to conduct a search for candidates for governor. It was a difficult position to fill. The specifications stated that the person chosen should have high-level experience in dealing with various government lending agencies, in the private investment field or in financing business enterprises. The person was also required to have, or quickly acquire, a broad knowledge of the rural electric program and to be sympathetic to the cooperative philosophy and objectives. Given the previous lack of interface between electric cooperatives and the private markets, these were almost mutually exclusive requirements. "It is not easy to find persons who have the right combination of qualifications for this position," Eric Nichol, vice president of Nelson, Nichol, told J. K. Smith.

Despite the difficulties, he presented the board with a list of 18 qualified candidates in March 1970 along with a list of 31 others who had been investigated. The personnel committee interviewed several candidates, but the ones the committee was most interested in were not interested in CFC. And there continued to be a high level of interest in a man who was not an official candidate: J. K. Smith.

Roger Oresman assisted the search committee in its interviews and activities. "Initially, the board was uncertain whether they needed someone with a strong financial background or a strong background in the cooperative movement," he said. "Those of us who had some financial background realized that what was really needed was someone with a strong cooperative background. You could hire the other kind of expertise. At that time, J. K. was president of CFC, and it was apparent almost from the beginning that he was the outstanding person in the group."

Smith for Governor

The issue was decided at a brief board meeting that followed the July 27–31 fly-in meetings. "The personnel committee has decided that J. K. Smith is the best prospect for the position of governor by reason of his administrative ability and experience, familiarity with the program and the people in it, compatibility with the REA policy makers and NRECA leadership, his demonstrated capability of meeting with financial people and grasping the intricacies of the money market, his acceptability to CFC applicants and approval of his performance on behalf of CFC by them," said Secretary-Treasurer Leland Leatherman. The board accepted the personnel committee's recommendation and appointed Smith governor.

Despite his commitment to CFC, the decision to accept the position of governor was apparently a difficult one for Smith. He had a comfortable position as a respected leader and head of the Kentucky Rural Electrice Cooperative Corporation, a statewide association. The new job was not without risk.

"J. K.'s willingness to leave his job as a statewide manager where he was safe and secure and to agree, for the sake of the program, to take on a new job that would be a real challenge

J. K. Smith
First Governor of CFC

J. K. Smith was an innovator, an entrepreneur and a great contributor to the rural electric program long before anyone ever thought about supplemental financing or the need to start an organization like CFC. In 1938, at age 21, he became the youngest person ever to serve as manager of a rural electric cooperative when he became manager of Fleming-Mason Rural Electric Cooperative Corporation in Flemingsburg, Kentucky. Under his leadership, it became one of the largest systems in the nation from the standpoint of area served. The cooperative was so successful that, given REA's prevailing attitude of the time that co-ops should be small, locally centered organizations, it was later divided into three systems.

Smith left Fleming-Mason in 1948 to help organize the Kentucky Rural Electric Cooperative Corporation, the first association of rural electric cooperatives in Kentucky. He served as its first executive manager. With his vision and guidance, the statewide organization developed numerous services for its 24 members, including an extensive manufacturing and sales program. The cooperative was the parent of three other companies, Consumers Credit, All-Electric Homes and Statewide RECC, a sales organization.

During his career, Smith was involved in numerous activities that affected every segment of the rural electric industry. He was an organizer and secretary-treasurer of East Kentucky Power Cooperative in Winchester, Kentucky. He was one of the founders and the was a measure of the character of the man. This new job had a chance of failure, and he knew it," said Long-Range Study Committee and CFC Board Member Owen Manning.

There was also a personal challenge. Harry Oswald, head of the Arkansas statewide association, sent Smith a letter with a copy to every rural electric system in the United States saying that the rural electrics had accepted the CFC concept on the basis of their faith in the work of J. K. Smith and his fellow

first president of the Central Area Data Processing Corporation in St. Louis, Missouri. He was also one of the organizers, a former director and past president of the Electric Research and Manufacturing Cooperative in Dyersburg, Tennessee. He spearheaded a drive in Kentucky to furnish obsolete and outgrown materials to construct a rural electric cooperative in Ecuador and made the original feasibility study to bring rural electrification to the Mekong Delta area of Thailand.

Throughout his career, Smith received many honors and awards. In 1966, he received special recognition from U.S. Secretary of State Dean Rusk for outstanding work in bringing rural electrification to other nations of the world. In 1967, the president of Ecuador presented him with the highest award that country can present to a citizen of another nation. He was the recipient of the Rural Electrification Administration's Distinguished Service Award in 1977. In 1979, he received the Co-op Career Award from the Cooperative League of the U.S.A. for his outstanding leadership and contributions to the cooperative movement.

One of Smith's greatest achievements was undoubtedly the work that led to the creation of CFC and its establishment as a viable institution. "J. K. had a presence. People respected him immediately," said his successor, Charles B. Gill. "There were people in our program who did not love him, but they did respect him. He was a real innovator, and he could visualize and create something. He created the statewide in Kentucky. He created the materials supply organization in Kentucky. He created CFC. His leadership at the time he was governor was very important."

CFC's board of directors recognized Smith's contributions by naming the CFC headquarters buildings, both in Georgetown and Herndon, after him and by recognizing him with the CFC Pioneer Award during the 50th anniversary celebration of rural electrification.

committee members. "Now," he wrote, "you have an obligation to all the cooperatives to make this plan work."

Smith had devoted considerable time and effort to the creation of CFC. "When things got down to decision time and he saw who the candidates were and he saw that the board wanted him, he agreed to do the job. I think he wanted to see to it that CFC did become a viable entity," Shesser said. "He put his best efforts into it."

Smith accepted the position effective July 31, 1970, but remained in Kentucky until November 1 to complete his work there. In August he came to Washington to consult with Shesser and Samenow about ongoing activities. He told them he was putting together some ideas about staffing. He had already recruited Paul Fox to be finance officer. Fox had no rural electric experience, but as vice president of finance for Arby's Roast Beef Franchising Company, he had experience with many of the financial issues facing CFC. Smith asked Samenow to become CFC's corporate counsel.

Shesser recalled that Smith then turned to him and said, "I'd like you to consider being loan officer."

Shesser looked at him and said, "J. K., I want you to know that I've been in the REA program for 20 years, and I have never done anything in loan processing. I have not done anything close to that type of work."

"Ira, I have every confidence you can do the job," Smith told him. Shesser served quite successfully as CFC's loan officer until his promotion to assistant to the governor in 1980.

Open for Business

CFC officially opened for business on September 15, 1970—almost three months ahead of the initial target date—in offices located at 500 12th Street, S. W., Suite 810, in Washington, D.C. The corporation executed a three-year sublease with Eastman Kodak Company for 3,125 square feet of office space in the new building. The location was only a block from REA's offices in the South Agriculture Building. The annual rent was $18,804, and CFC spent an additional $14,000 on furnishings and $1,500 on structural work.

Although the lease was for three years, one demonstration of CFC's early success was that it quickly outgrew the 12th Street facility. The operation moved into new offices at 300 7th Street less than a year later on June 1, 1971, tripling its space.

There were some people in the program who said CFC only needed three employees: a secretary to answer the telephone, J. K. Smith to approve loans and an accountant to write the checks. "That attitude prevailed for quite some time," said Gill. "Only after REA began to diminish its role and CFC picked up a greater role did people begin to realize what sort of organization they had here."

Smith had a philosophy of keeping the staff lean. The new employees did whatever had to be done. "I was the finance officer," Fox said. "I was also the administrative officer and the controller, and occasionally I even cleaned the floors when the cleaning people didn't do the job too well."

A number of CFC's initial employees and consultants had years of experience in the rural electric program. "That was very sensible for the start-up phase," said John Jay List, senior vice president and general counsel of CFC. "Basically, what CFC had to do was to understand REA and get along with them. The way to do that was to utilize people who knew their lending approaches, their documents and that sort of thing."

The first official employee was Lee Kalbacher, who learned about the new organization while working at NRECA. She actually started work as secretary on July 1, 1970, to assist with the administrative details of the organization process. In addition to Governor Smith, Finance Officer Fox, Loan Officer Shesser and Corporate Counsel Samenow, CFC engaged Milton A. Chase as consulting engineer and Robert Beall as staff consultant. Herman Gelman soon joined as senior accountant to write the checks. With that, CFC had the core of people it needed to begin operations.

Establishing CFC's Financial Structure

By the November 1970 board meeting, CFC had 803 applications for membership. Of those, 763, including 736 Class A members (distribution systems) and 27 Class B members (power supply systems), had subscribed to CTCs. The remaining 40 applicants for Class C and D membership, including statewides, service organizations and NRECA, were not required to subscribe to CTCs. The subscription commitments totaled $115.2 million, and CFC had already received $46.1 million.

There were no loans yet because CFC was still working out the details of its agreements with REA. Thus, CFC became an investor before it became a lender. "We invested the funds in a limited scope of securities, principally certificates of deposit of banks that had capital and surplus of $50 million or more," Fox said. "Government-guaranteed or federal agency debt was also a permitted investment. Our objective, of course, was to earn the best return we could and also to spread our business among banks with which we might be developing relationships in the future."

Member investments in Capital Term Certificates provided an important equity base for CFC's collateral trust bonds and other debt instruments. Pictured: Paul Fox, finance officer, reviewing one of the first CTCs.

The basic loan policy outlined in the Invitation to Subscribe to Capital Term Certificates said that CFC would allocate its resources to meet both short-term and emergency construction needs, as well as the long-term requirements of the distribution systems and G&Ts. Systems could apply for 100 percent CFC loans for special purposes, as well as concurrent loans (originally called mixed loans) involving both REA and CFC. The policy established the times interest earned ratio (TIER) as a criteria for loans to distribution systems in addition to the PRR. TIER measures a system's ability to earn annual revenues to cover interest payments on its debt. Systems with a TIER of less than 1.5 or a PRR greater than 9.0 would not normally be considered eligible for a loan. In June 1971, a debt service coverage (DSC) ratio of 1.25 was added to the loan criteria. DSC is a measure of a corporation's ability to earn annual revenues to cover all debt components, both interest and principal. This further strengthened the collateral supporting CFC's securities.

CFC was able to offer a short-term loan program before the completion of documents for the concurrent loan program. Governor Smith approved the first loan, $500,000 for a period of six months, to Warren Rural Electric Cooperative Corporation of Bowling Green, Kentucky, on December 8, 1970. The purpose of the loan was to fund plant additions pending the approval of a long-term loan by REA. The loan, which was not secured, carried a variable interest rate based on the prime rate. The initial rate was 7 percent.

At the time, the introduction of short-term financing to the cooperatives was quite innovative. "It was something they had

CFC loan and financing documents established minimum times interest earned ratio (TIER) and debt service coverage (DSC) ratio requirements to ensure the creditworthiness of its member systems.

$$\text{TIER} = \frac{\text{Patronage Capital and Margins} + \text{Interest on Long-Term Debt}}{\text{Interest on Long-Term Debt}}$$

For a distribution borrower, the REA-CFC common mortgage and CFC's initial indenture required an average TIER of 1.5 for the two highest of the last three years. For a power supply borrower, the requirement was an average TIER of 1.0 for the two highest of the last three years.

$$\text{DSC} = \frac{\text{Patronage Capital and Margins} + \text{Interest on Long-Term Debt} + \text{Depreciation and Amortization Expense}}{\text{Sum of All of the Year's Required Payments of Principal and Interest on Long-Term Debt}}$$

For a distribution borrower, the REA-CFC common mortgage and CFC's initial indenture required an average DSC of 1.25 for the two highest of the last three years. For a power supply borrower, the requirement was an average DSC of 1.00 for the two highest of the last three years.

Note: CFC established a new indenture in February 1994 that contains no TIER requirement, an equity requirement of 20 percent and a modified DSC requirement for distribution systems of 1.35 for the average of the two highest of the last three years. The indenture does not address requirements for G&T systems because CFC does not anticipate pledging G&T collateral as security. Collateral pledged under the initial indenture must continue to meet the requirements of that indenture.

$$\text{Modified DSC} = \frac{\text{Patronage Capital and Operating Margins} + \text{Interest on Long-Term Debt} + \text{Depreciation and Amortization Expense} + \text{Cash Patronage Capital Retirements Received}}{\text{Sum of All of the Year's Required Payments of Principal and Interest on Long-Term Debt}}$$

not really used," Shesser said. "They were restricted by REA, and if they wanted to do any other borrowing, they had to get clearance. Introducing the line-of-credit program very early on, even before we began long-term lending, opened up a whole new era."

Implementing the short-term loan program at that time was another demonstration of Smith's creativity and innovation. "My own initial reaction was, why do the co-ops need this? They

REA Administrator David A. Hamil announced the first two long-term, secured loans to be made concurrently by REA and CFC on February 16, 1971, during the NRECA annual meeting. Pictured: CFC Governor J. K. Smith **(left)** ***and Administrator Hamil*** **(right).**

have never had it," Shesser said. "J. K. was convinced that if it was made available, it would become a very useful tool." That turned out to be true. Over time, even systems that did not anticipate a need to use a line of credit found it useful to maintain one as a reserve against an unexpected event.

REA and CFC's First Concurrent Loans

By the 1971 annual meeting in Dallas, Texas, everything was in place for CFC to begin its long-term concurrent loan program.

On February 16, during the NRECA annual meeting, Administrator Hamil announced the first two long-term, secured loans to be made concurrently by REA and CFC. One was a $1 million loan to North Arkansas Electric Cooperative, Inc., of Salem, Arkansas. The REA portion was $900,000 whereas CFC's portion was $100,000. The other loan was to Washington County Electric Membership Corporation in Sandersville, Georgia. The total amount of $389,000 was split between REA, $350,000, and CFC, $39,000. The CFC interest rate of 7.5 percent was reduced to7 percent after CFC's first bond issue.

By the end of the fiscal year on June 30, CFC had approved nine short-term loans totaling $3.5 million and 30 concurrent loans totaling $12.8 million.

Overall, lending operations during the early years were slow. REA had agreed that all loan applications received as of December 18, 1970, would be processed as 100 percent REA loans. "REA was very sympathetic," Shesser said. "They realized that this was a big step that the rural electrics were taking, and they made every effort to have the transition run as smoothly and painlessly as possible."

The 1971 annual meeting was important for another reason. The membership of CFC elected directors for the first time. Under normal procedures, candidates for director positions should have been elected at the 1970 district meetings. However, the board was unable to complete all the arrangements before the meetings ended. The 1970 district elections were the first and only elections in CFC's history conducted by mail.

The first national election was held without difficulty at CFC's second annual meeting on February 16, 1971, in Dallas.

CFC Serves the G&Ts

One issue that plagued CFC throughout this period was the question of what types of services it could provide to the G&Ts. The type of financing needed by the distribution systems—large numbers of loans for relatively small amounts—was very different from that needed by the G&Ts—a small number of loans for very large amounts. The growing need for financing for generation was one of the factors that led to the creation of CFC, but as a small, untried organization, its ability to finance large projects was limited. The initial loan policy in the Invitation to Subscribe to Capital Term Certificates said that because of the widely varying circumstances among power supply systems, loans to G&Ts would be considered on a case-by-case basis.

"When we first announced our basic loan programs, CFC was criticized by the power supply systems because we did not have any type of program designed for them," Smith said. "We did not spell out directly how a power supply system could borrow any substantial amount of capital from CFC. The G&Ts realized that the time would come when they would need substantial amounts of supplemental capital. They wanted an alternative source, in addition to REA, to borrow from, if the need arose."

Smith attended a national meeting of power supply systems held in Louisville, Kentucky, on July 23, 1970, to discuss the problem. Forty G&T systems were represented. Smith promised to give further consideration to power supply system capital needs and means of adequately meeting them. It was the beginning of a dialogue that resulted in several programs that provided substantial savings and benefits to the G&T systems.

The G&Ts organized a committee, including E. V. Lewis, a CFC board member and manager of Central Electric Power Cooperative, Inc., in Columbia, South Carolina; George Taylor, manager of South Mississippi Electric Power Association in Hattiesburg, Mississippi; Hugh Spurlock, manager of East Kentucky RECC (later East Kentucky Power Cooperative) in Winchester, Kentucky; Harry Oswald, manager of the Arkansas statewide and the fledgling Arkansas Electric Cooperative Corporation in Little Rock, Arkansas; Wendell Garwood, executive manager of Tri-State G&T Association in Denver, Colorado; and Andy Freeman, manager of Minnkota Power Cooperative in Grand Forks, North Dakota, to plan for a second G&T meeting in November. The group officially organized as the G&T Managers Association early in 1971 for the purpose of identifying and resolving their financing and other problems.

The short-term loan program was the first to offer something to the power supply systems. It gave them a way to finance all the work that needed to be done on a project before REA could approve a loan and advance funds. "This received immediate support. We termed this a revolving loan, and it was an entirely new concept for the rural electric systems," Smith said.

Valuable as it was, it didn't address the systems' long-term needs, and CFC did not yet have anything to offer in that area. Representatives from the newly formed G&T Managers Association met with the CFC board in March 1971. The systems foresaw a need for $500 million over the next 18 months. REA said it could only help in a small way and recommended that they develop other sources of financing. "CFC has a definite program for helping the distribution cooperatives and has established guidelines for them, but you have not done the same for the power supply systems," Spurlock, chairman of the Association, told the board.

The group had explored the possibility of working jointly with municipal power systems to develop generating systems and somehow gain access to tax-exempt financing. That approach appeared to have limited application. The lack of equity presented a problem with other approaches.

"What the power supply systems need most is a single source of long-term financing at the lowest cost available," Garwood said. "We hope that CFC will show us a way to meet this need."

CFC had no immediate answers, but Katzin said that if the power supply systems would select CFC as their financial coordinator, it should be possible to arrange private market financing for them when required. Subsequently, Spurlock; John Madgett, manager of Dairyland Power Cooperative in La Crosse, Wisconsin; and Truman Green, manager of Central Electric Power Cooperative in Jefferson City, Missouri, were appointed by the Association to work with CFC, and East Kentucky RECC formally asked CFC to serve as its financial coordinator.

30–10–60 Plan to Finance Generation

By August, there was a plan. Over the next year, CFC worked with REA and the G&T Managers Association to develop the 30–10–60 program as a means of raising capital for the power supply systems. Under this concept, REA would make a loan for 30 percent of the cost of a project, and CFC would lend 10 percent. The CFC and REA loans would be subordinated for use as equity capital. CFC would help the system raise the remaining debt from another source, likely to be insurance companies.

A rather convoluted lending structure was devised to achieve this. The REA and CFC loans were actually made to the power supply systems' members, who reloaned the funds to the G&T. The REA and CFC loans were secured by a first mortgage on the distribution system property, and the members were secured by a subordinated mortgage on the power supply system. "This was a complex plan, but it worked," Smith said.

CFC also began work on a bulk power supply financing program that would allow the G&Ts to finance projects with funds from REA and CFC without having to borrow directly from the private market. The plan called for REA to provide the first 30–40 percent of the funding for the project, which would be treated as equity. The borrower would provide additional equity through the investment in Loan CTCs required as a condition of each CFC loan. CFC would provide the remaining debt by issuing bonds under a separate bulk power supply indenture. CFC would also make short-term construction loans to support the project. The proposal was complex, and a significant amount of time was needed to develop documents and negotiate various issues with REA. The 30–10–60 plan provided an interim source of funding for G&T systems with immediate needs for capacity additions.

East Kentucky Power Cooperative's Charleston Bottoms Station was the only facility financed under the 30–10–60 program developed by CFC and REA to fund construction of generating plants. Pictured,* left to right: *CFC Governor J. K. Smith, East Kentucky Power Cooperative President Alex B. Veech, CFC Loan Officer Ira Shesser, East Kentucky Power Cooperative General Manager Hugh Spurlock and REA Administrator David A. Hamil.

REA and CFC agreed to make loans to two systems under the 30–10–60 program. CFC approved its first loans to the members of Associated Electric Cooperative, Inc., on June 12, 1972, for a 600-megawatt coal-fired unit at the New Madrid plant. The total cost of the project was estimated at $240.6 million with CFC providing $24.1 million, REA $72.2 million and other lenders $144.4 million at an interest rate of 8.375 percent.

During the course of the negotiations over the loan documentation, Associated became concerned about the remedies available to the private lender in the event of default. In an effort to counteract those concerns, CFC agreed to a ceiling on the interest rate. As interest rates rose, that commitment created some tense moments for the CFC staff. "We were never concerned that we couldn't raise the amounts, but we were afraid that our cost of funds would exceed the interest rate ceiling," Chase said.

Due to changes in the REA program, the funds were never actually advanced on the Associated loans.

On June 29, 1972, CFC loans totaling $13.2 million went to

16 of the 18 members of East Kentucky RECC to help finance the 300-megawatt Charleston Bottoms Station. The total cost of the project was $125 million, with REA providing $37.5 million and another lender $75 million at an interest rate of 8.5 percent. East Kentucky proceeded with its financing under the plan and remains the only rural electric system with a 30–10–60 financing in place.

"We knew this approach wouldn't meet the needs over the long-term," Smith said. "It was a stop-gap measure needed at the time. It enabled us to meet a current requirement on an immediate basis." It also demonstrated CFC's commitment—and ability—to meet the needs of its power supply members.

Approaching the Markets

While Shesser, Beall and others were working to get the loan programs going, Paul Fox was establishing the financing program. With the CTC investments coming in, CFC did not need to go to the market for funds immediately, but everyone recognized that it would take time to prepare for that eventuality. The first two priorities were to establish programs for raising different maturities of debt and to develop commercial banking relationships.

"We knew we would need short-term debt, because we would be making short-term loans in the near future. We certainly knew we would need long-term debt because our principal lending program would eventually be making long-term, 35-year loans to our member cooperatives," Fox said.

Long-range plans and strategies occupied most of the board discussion on financing at that time. "Should we go out for debt for as long as 35 years? Should we use one investment banker? How should we structure our indenture? Throughout this entire early planning process, we were trying to figure out how to describe all of these issues to our board members in a clear-cut way so that they could understand, discuss and criticize our approach," Fox said. "We very much wanted to involve the board in all our planning."

By all accounts, Katzin was a master at demystifying Wall Street, educating the board and making them comfortable with the decisions they were responsible for making. "There has been a great agrarian tradition in this country for the last century of the farm against Wall Street. Wall Street was the evil money power," Katzin said. "It was the money lenders who foreclosed on the mortgages and the mysterious bankers who controlled the prices. There has always been and always will be a deep suspicion between Main Street and Wall Street."

Katzin overcame that, not only with the CFC board but also with others in the program, by teaching them the language and explaining the secrets. "Every craft, every art, every profession has its own language," he said. "Wall Street has a language. Once you realize that it is simply words that they use to describe certain events or certain transactions, it is no more mysterious than anything else."

CFC's financial strategy evolved over time. "Generally our thoughts were that we were going to need to finance our members' needs, and we were going to have to go to the private markets. We were not yet sure how we were going to do that," Fox said. "CFC was a totally unique corporation. There was no textbook to refer to, no experts who had done this kind of financing many times before."

Katzin and his colleagues at Kuhn, Loeb along with Oresman and his associates at Milbank helped CFC's staff and board devise an approach that was very successful from the beginning. The cornerstone of the approach was the CTC. "It really was a brilliant concept, and this instrument is one of the most important reasons for the favorable image that CFC has acquired in the financial community," Fox said.

The willingness of the rural electric systems to make investments in CFC over a 15-year period, investments that carried a term of 50 years, demonstrated that CFC had strong long-term support for its programs. While the consultants worked on the indenture that would allow CFC to raise debt based on that support, the CFC staff, along with Kuhn, Loeb, started to tell the financial community just who CFC was.

CFC continued to work through Kuhn, Loeb in developing investment banking relationships but decided to develop commercial bank relationships independently. "Quite honestly, I didn't think it was appropriate or proper for me to begin contacting other investment bankers, even for information, at that early stage of CFC's development," Fox said. "I think we owed at least that much to Jerry Katzin, as well as an opportunity to see if his ideas could bear fruit."

Developing Relationships with Commercial Banks

Developing a group of banks who would understand CFC and offer lines of credit to CFC was an important activity, even though CFC would not actually need the lines until loan activity increased. CFC could have selected a lead bank to serve as its banker. Under that approach, all of CFC's dealings would be with the lead bank. It would execute a loan agreement with CFC and evaluate CFC's credit. In turn, it would syndicate the loan to as many other banks around the country as it needed to meet the obligation. "That is the easy way, but it puts you in the hands of that one bank," Katzin said. "The bank could become disenchanted with you or choose not to be too competitive, or for any other reason, may not be there to support you when you need them."

Instead, CFC decided to act as its own syndicator and to arrange the credit to CFC on an individual basis. This approach required each of the commercial banks to become familiar with CFC rather than relying on a lead bank for its judgment and knowledge of the credit. The strategy meant more work for the CFC staff, but it was a very important element in building the financial structure of CFC. Within a few years, CFC had developed relationships with the major money center banks in the United States as well as important regional banks and had established relationships with Canadian, British, German and French banks.

Katzin credits Fox with developing the commercial bank program. "We didn't foresee the reliance on bank financing. It has given CFC a tremendous reach, extending to all parts of the country and, indeed, worldwide. It has made possible types of funding that are much farther advanced than our initial bond issue," he said.

The decision to develop the contacts in-house helped to position CFC in establishing long-term relationships. "We knew we would be dealing with a large number of banks eventually," Fox said. "By choosing not to use an agent, we were able to present CFC as a living organism, rather than just another credit. CFC continues to operate on this basis with a lot of its financing. If we had this decision to make again, I believe we would do it the same way."

In the early days, a typical meeting with a bank would involve Fox, Smith (if he was available) and perhaps a board member from the area in which the bank was located. Fox always tried to see the most senior person at the bank to establish that CFC was going to be an important financial organization in the future. They would describe CFC and why it was created, its relationship with REA and the government and the very strong credit record of the member systems. "We stressed the romance or the drama of CFC," he said. "It was new. It was a very unique organization. There was really nothing like CFC in 1970."

CFC's presentation emphasized the creditworthiness of CFC's member-borrowers and the quality of management at CFC itself. "The fact that billions of dollars had been loaned to the rural electric systems by REA and less than one-tenth of 1 percent of the amount of these loans had ever been written off impressed them," Smith said.

Many of CFC's senior staff members had extensive experience in rural electrification. "Some banks were fascinated with the degree of knowledge that we had about rural electrification and of CFC itself," Fox said. "It is surprising, but there are representatives who talk about their companies but just don't know too many of the details. Our approach of putting the whole story on the table was well accepted."

The board of directors was an important selling point. "The board was indeed from the very community that we represented and so had the program of CFC very much at heart," Fox said. The board also worked very hard. "Many of the banks were amazed when we said that a typical board meeting at CFC lasts two and a half to three days. Normally, a board meeting at a bank lasts two or three hours."

The Asset of Member Diversity

CFC's diversity was another asset. "The basic strength of CFC, apart from the quality of management, which has been very high with both governors, and the quality of the staff, which is superior, together with the dedication of everyone in the program, has been that it represents a unification of a vastly diversified group of cooperatives into a single instrument in the marketplace," Katzin said.

CFC's membership comprised a diverse community of borrowers both in terms of numbers and geographical diversity. "The wonderful thing is not only the diversity of size, the diversity of membership, the diversity of miles of line, but the diversity of geography," Fox said. "That is what makes this system low risk.

If there's an earthquake in Oregon, the sun is shining in Florida."

CFC emphasized that strength by inviting the loan officers of each bank to attend the CFC annual meeting held each year in conjunction with the NRECA annual meeting. It provided an opportunity to talk to all of them at once, and more importantly, it let them see the support behind CFC in the form of the thousands of people attending the meeting. CFC hosted a financial forum during the meeting at which Fox, in a remarkable demonstration of memory, recognized each bank representative by name. The performance always brought the house down.

Fox frequently included Katzin or one of his associates in his early visits with the commercial banks. "I did this for two reasons: first, to have a check on my performance in describing CFC, and second, frankly, to exploit the value of the Kuhn, Loeb name, which was very valuable in both the banking community and the utility community," Fox said. Although CFC was an unknown quantity, Kuhn, Loeb had a flawless reputation and was well-respected throughout the financial community.

In November 1970, Smith and Fox were invited to lunch by Joseph Barr, chairman of American Security and Trust Company, and one of his loan officers. (Barr served as undersecretary of the Treasury Department and, briefly, as secretary of the Treasury Department during the Johnson administration.) "It was a very elegant lunch at the Madison Hotel," Fox said. Smith told them about the Long-Range Study Committee, and Fox discussed the prospects for CFC.

At the end of the lunch, Barr said, "Well, gentlemen, you have a $5 million line of credit from American Security and Trust Company."

The two were surprised to have one offered so early in their efforts. "We recognized that it demonstrated a positive reaction to our enthusiasm, to the solidity of our idea, to the romance of it and to the dramatic nature," Fox said. "We didn't take it, incidentally, because we didn't need it yet."

American Security was very helpful to CFC in other ways. It frequently held seminars for other banks throughout the country and invited CFC to speak at the seminars. "That was wonderful for us because it introduced us to at least 40 other banks with whom we later established relationships," Fox said.

When CFC did ask for lines of credit, the response was good. "Bankers generally take a very conservative view in providing bank lines of credit, but after we had finished our presentation, they would ask if we were sure we didn't want a larger line of credit," Smith said. "We never had a bank turn down our request."

However, the banks did have questions. "Bankers pride themselves on being well-informed, but many whom we visited had never heard of the rural electrification program," Smith said. "A few wanted to know how we could pay our debts and stay in business if we didn't make a profit. However, once they understood the unique relationship between CFC and its members and the true meaning of member ownership, these questions disappeared. We had to sell the cooperative concept."

The banks had also had good experiences with rural areas and with electric utilities. They could see the business potential of the new organization. Some wanted to match their line-of-credit commitment with the amount of money that would be loaned in their respective states. "We were able to convince them that their security would be stronger if they were a part of CFC's overall credit line."

The CFC board approved the first two lines of credit during the March 9–10, 1971, board meeting: a $1 million line with American Security and Trust Company and a $2 million line with Riggs National Bank.

Series A— The First Collateral Trust Bond Issue

The strategy CFC developed for approaching the long-term market was to use its equity—the investments in Subscription CTCs—to build a portfolio of loans. "When we reached a sufficient amount in loans to provide collateral for a bond issue in the public market, we would sell our first issue and raise capital with that collateral, thereby freeing up that money to relend," Katzin said. "We would, in effect, be revolving our equity and using these funds to make new loans that would then form the collateral for a new financing."

The CFC staff and consultants worked hard to get CFC ready to issue collateral trust bonds. "One of the very first things we had to do was to draft provisions of an indenture that would give us adequate security for our collateral trust bonds," said Louis Roland, by then CFC corporate counsel.

The indenture, also known as a deed of trust, set forth the agreement as to the respective rights and duties of CFC, the bondholders and the trustee under which the bonds would be issued. The collateral trust bonds issued under the indenture were backed by the corporation's pledged assets, a portfolio of mortgages deposited with the trustee.

"The CFC indenture of pooling a lot of obligations and diversifying the risk of loss from any one obligation was the forerunner for the mortgage-backed securities program that is so prevalent in the marketplace today," said Robert R. Edmiston, managing director of Lehman Brothers, the successor to Kuhn, Loeb. "When CFC did its first collateral trust bond issue, there was no mortgage-backed securities market. The first were sold in 1974."

The indenture was as important to CFC in terms of borrowing money as the common mortgage was for lending money. It was not an easy document to write. "We had three or four meetings a week at CFC while we were working this out," Roland said. "Our telephone bills were astronomical. We would sometimes have two-and-a-half-hour conversations with our general counsel's office in New York on one provision. A tremendous amount of time was spent in working out the indenture provisions."

The indenture had to be structured to reflect the fact that CFC was a cooperative. "As a cooperative, we required flexibility in the provisions of the indenture that was not present in an ordinary indenture," Roland said. "This indenture had provisions that would govern CFC until the bonds were paid off. It would provide a type of constitution that CFC could not violate and that would govern all our future operations. Amending substantive provisions of the indenture would be a most difficult process. We could not afford to make mistakes."

The indenture established standards for the CFC mortgages that could go into the collateral portfolio. CFC could also provide cash and investments as collateral. The provisions of the original common mortgage with REA were used as the general basis for the type of security that would be pledgeable under the indenture. CFC offered one dollar of mortgage collateral for every dollar borrowed. It was suggested at one point that it might be necessary to overcollateralize by offering, for example, collateral equal to 120 percent of the funds borrowed. "We decided to cut this as tight as we could and did not elect to follow that course," Katzin said.

The loan portfolio was evergreen. That is, if one piece of collateral failed to meet the required standards, CFC could replace it with collateral that did meet the standards, so that the bonds were always supported by a pool of qualified collateral.

There were also two important restrictions. There was a limit on the amount of notes that could be pledged by CFC to the indenture trustee from any single member, and only 25 percent of CFC's portfolio could be used for G&T notes.

Member Performance Supports CFC's Indenture

Another unusual aspect of CFC's indenture was that CFC itself was not required to meet any specific financial ratios: only the borrowers whose mortgages were pledged as collateral were required to maintain a specific level of performance. However, CFC was required to make mandatory sinking fund payments into a debt service reserve each year to ensure that funds were available to redeem the bonds.

Once these concepts began to take form, CFC started to discuss them with the rating agencies. "The tactic of going to see the rating bureaus early, when we only had general ideas about the form of the indenture, was brilliant," Fox said. "By taking that approach, we created a receptive attitude on the part of the rating agencies." Credit for that idea, like so many of the others that made the structure of CFC successful, goes to Katzin.

Two rating analysts in particular were involved from the very beginning. Al Esokait was the senior analyst at Moody's Investors Service at the time and was respected throughout the industry as the dean of the rating business. "He was intrigued by the CFC idea, and he took a lot of time to help us structure our indenture in such a way that our bonds would receive at least a single-A from Moody's," Fox said. Russell Fraser, who was in charge of corporate ratings for Standard & Poor's Corporation at the time, was also intrigued by CFC.

"Each of them was very helpful," Katzin said. "They took the time to understand CFC and to make sure that their staffs gave it the attention and study required by this new creature in order to properly appraise its creditworthiness."

A-Rated Series A

When the time came, CFC's initial bond issue did receive an A rating from both agencies. That was important for three reasons. First, that rating made CFC bonds a legal investment for insurance companies in most states. Second, the rating denoted a high credit quality. Finally, it was attractive to potential investors because of the possibility of a future upgrade to double-A, which would enhance the value of the bonds.

"We were told then that one reason we did not get a double-A rating initially was that, while our standards entitled us to that quality, the rating agencies naturally had to see how we were going to function," Katzin said. "They needed to have some years of experience with us and with the board, as well as with the management performance of the cooperatives, before they could honestly give us a double-A rating."

CFC knew it was going to issue bonds. The question was when. Katzin suggested as early as the spring of 1971 that CFC consider going into the bond market to raise capital for CFC, even though there was no immediate need for the funds. His reasoning was that, with rates on 35-year A-rated debt at 7.875 to 8 percent, the timing was good from a rate viewpoint. Also, CFC wanted its name to be known in the investment community. A bond issue was a sure way to gain recognition. And CFC was making commitments to advance funds to members in the next 6 to 24 months.

However, aside from the fact that CFC had no need or use for the funds, the bond indenture was a very complex document that was not yet ready. It took more than a year to develop the indenture, which exceeded 170 pages in length, and related documentation for SEC registration, and there were other issues that had to be addressed.

Joseph Payne, vice president of the Chase Manhattan Bank, met with the board in February 1972 to discuss the role of the trustee under the indenture. He said that Chase Manhattan had the largest corporate trust business in the United States. The trustee, he explained, had three functions. First, it assisted in the development of the basic indenture. Then, it helped prepare the bonds or securities to be issued. Finally, the trustee serviced the bonds by maintaining records, making interest and principal payments, transferring and exchanging the bonds and responding to inquiries. Should an issue go into default, the trustee would assume the administration of the trust.

Payne said that his company would be interested in working with CFC as trustee on its bond issues, and the board subsequently approved the Chase Manhattan Bank of New York as trustee under the indenture for collateral trust bonds.

The board also designated Kuhn, Loeb as the managing underwriter for CFC's first issue. They put together a syndicate that included a substantial cross section of the underwriting firms in the country. "It was our custom to invite all of the major underwriters to join us," Katzin said. "Usually, the underwriters just fall into line when a new name is introduced. However, we met with some resistance among the principal underwriters. This was for the very simple reason, which most of them stated and some did not have to state, that these were all houses that had major participation with the private utility industry. They did not want to risk the hazard of damaging those relationships where they were the principal managing underwriters by joining in this cooperative venture."

The concern was well-founded. Kuhn, Loeb was never a major force in public utility financing after its association with CFC. "There was a much greater degree of enmity between investor-owned utilities and cooperatively owned utilities going back into the 1960s and 1970s than there is today," Edmiston said. "It is fair to say that CFC's association with Kuhn, Loeb & Co. at the time cost Kuhn, Loeb & Co. quite a bit of business with investor-owned utilities." It is one measure of the firm's integrity that it did not allow that to affect its relationship with CFC. "There was an enormous commitment on the part of the firm to National Rural Utilities." Fortunately, the animosity between cooperatives and investor-owned utilities subsequently abated. "I have not seen or heard of any concern on the part of an investor-owned utility with respect to Lehman's affiliation with CFC."

By the fall of 1972, CFC had Subscription CTCs totaling almost $125 million and approved loans to members totaling almost $50 million. Although refinements were made until the very end, the basic documentation was complete. The rating agencies were well-informed. It was time to issue CFC's first series of bonds—Series A. CFC filed a registration statement with the SEC to make an initial public offering of $50 million of 35-year collateral trust bonds on October 12.

Educating Potential Investors

In marketing Series A, CFC had to educate several groups of people. "The first, and a very important group, was the staff at Kuhn, Loeb," Fox said. "These were the people who would be involved in marketing the CFC bond issues. If they didn't understand what we were about, we were sunk."

The next step was for Kuhn, Loeb to arrange visits with the other investment bankers who were going to be underwriting and selling the issue. "Again, if they didn't understand us, we felt that we could not be successful. We were trying to get them to see what we were bringing to the market," Fox said.

CFC then met with investor groups, including portfolio managers and decision makers, in major cities. "CFC staff spent three to four weeks visiting with potential investors," Smith said. "They were interested in knowing more about CFC because we were a new entity. They were not familiar with the type of securities we were bringing to the market, but they were willing to listen. We would tell them who we were, what we were about, what we were trying to do and how our program would work. They, in turn, had many questions concerning rural electrification, the ratings on our bonds and our method of operations."

The investors needed to understand the financial structure of CFC, but more importantly, they needed to understand the structure and history of CFC's members, the rural electric cooperatives, because they would be looking through CFC to the member systems for their basic security. There was some confusion over CFC's dual nature. The rating agencies rated CFC as a utility, but it operated as a finance company.

CFC's Greatest Strength: Strong Members

As in the presentations to commercial banks, CFC emphasized the strong credit history of the rural electric program and the moral character of rural people. Vincent Slatt, who succeeded Smith as president of CFC, had the job of explaining distribution systems. "One of the things that got their interest was the fact that our members read their own meters," Slatt said. "I explained that in order to keep expenses down, the members of the cooperatives read their own meters. The bankers could not believe that." The point was well taken, however. People who could be trusted to read their own meters could probably be trusted to pay their other debts.

The only significant negative was raised by one of the investment banking firms, which questioned how CFC could sell securities at a higher interest rate than the rate on outstanding loans. By blending the cost of the bonds with the cost of the CTCs, CFC could easily support an interest rate lower than that on the bonds. However, the concern of the investment firm was that if there were a default by CFC and the institution who held the obligation had to look to the collateral, that collateral would not meet the face amount of the obligation. "That was a cold, hard fact," Katzin said. The firm dropped out of the syndicate after the first issue, but as CFC's reputation and experience grew, it later returned.

CFC staff also spent time educating the CFC membership. "We tried to explain what it all meant in basic terms, what our obligations were, what their obligations were, what restrictions it might impose on them," Fox said.

Overall, CFC's reception in the marketplace was positive. "The enthusiasm we showed when we went to the market for the first time was appealing to both the underwriters and the investors. We convinced them we were sincere in what we were telling them, and we found them quite receptive," Smith said.

The Series A bonds were sold on December 7, 1972. The interest rate was 7.4 percent, and the bonds were priced at 99.625 to yield 7.43 percent. "That was a highly favorable rate and was comparable to what a single-A investor-owned utility bond would have sold for," Fox said. "The other measure of its favorability was in the aftermarket. After the bonds are sold to investors by the underwriting syndicate, the price of the bonds would be erratic if the original interest rate was inappropriate. The narrow range of aftermarket trading indicated that it was a very appropriately priced issue."

In a December 8, 1972, report on the issue, the *Wall Street Journal* said, "The single-A rated securities were purchased mainly by institutional investors, who were said to be impressed by the unusually lengthy 10-year period of refunding protection and by the strong credit record of the various rural electric utility systems belonging to National Rural. These two favorable factors offset the normal buyer resistance to any untested new market instrument, some analysts said."

CFC priced its initial long-term loans, which were funded out of equity, at 7.5 percent, based on the assumption that long-term funds could be raised at 8.5 percent. When Series A sold at a much more favorable rate, the board repriced all of the 7.5 percent loans to 7 percent during the January 1973 meeting. "This was a way of keeping faith with the members that CFC would operate on a cost basis," said Gill, then CFC's borrowers operations officer.

By the end of 1972, CFC had established a lending program.

It had established a financing program and had successfully marketed bonds.

"We were on our way," Smith said.

PART 3

CFC PROVES ITS WORTH

CFC established its value to the rural electric program during the first five months of 1973 after an action by the Nixon administration that effectively terminated the REA program. CFC advanced funds to meet critical financial needs while program leaders negotiated a new approach to rural electric financing. CFC helped its member systems establish sound financial management practices through services and investment opportunities such as the CFC commercial paper program. Through its support of the Capital Credits Study Committee and the Power Supply Study Committee, CFC focused attention on critical program issues. CFC provided leadership by making its first retirement of patronage capital in February 1980. By effecting an orderly transition of power on the board of directors and in the governorship, it demonstrated that it was capable of evolving to meet the ongoing needs of its member systems.

Highlights

- *CFC approved $44.6 million in short-term loans between January and April 1973 to meet the immediate needs of its members during the Nixon-imposed impoundment of REA funds.*

- *In November 1973, CFC established the Capital Credits Study Committee.*

- *CFC held its first Power Supply Symposium in December 1973.*

- *CFC began to issue commercial paper to investors through Goldman Sachs in May 1974 and directly to members through the Money Desk in September 1974.*

- *CFC reached $1 billion in loan commitments in 1974.*

- *CFC purchased property for CFC Square in Georgetown in May 1975, occupied the North Building in October 1975 and dedicated the Smith Building in December 1978.*

- *The CFC board authorized CFC's participation with NRECA in the Power Supply Study Committee in May 1977.*

- *CFC issued its first guarantees for $108,985,000 for pollution control equipment for nine G&Ts in November 1977.*

- *CFC reached $1 billion in loans outstanding in 1979.*

- *Charles B. Gill succeeded J. K. Smith as governor of CFC on September 1, 1979.*

- *CFC made its first distribution of patronage capital in February 1980 when it mailed checks totaling $2.6 million to member systems for patronage capital allocated during fiscal year 1973.*

Friday, December 29, 1972, was a typical winter day. Skies were overcast, and a few flakes of snow were falling. Business was slow, and the CFC staff was looking forward to an extra day off for the New Year's celebration. People were feeling good, not only because of the season, but also because things were going so well. CFC's loan program was well underway, and the first CFC issue of collateral trust bonds three weeks earlier had been highly successful. CFC had attracted a strong membership base, and loan volume was growing.

In midafternoon, Clyde Denton, CFC's information and member relations officer, was thumbing through some recent press releases from the Department of Agriculture when one caught his eye. He read it twice, then hurried to J. K. Smith's office with totally unexpected news.

The release said that the REA loan program was being converted to insured and guaranteed loan programs under the authority of the Consolidated Farm and Rural Development Act as amended by the Rural Development Act of 1972. "This change will eliminate direct... federal loans and substitute credit from private sources at interest rates that are more in line with the cost of money in today's market," it said. The action was effective January 1, 1973.

Program leaders, including REA Administrator David Hamil, who was home in Colorado for the holidays, were caught unaware. There had been no prior discussion or consultation with rural electric representatives. Many systems viewed the action as a direct attack on rural electrification. Although that was certainly the result, the change was apparently intended to eliminate the impact of the REA loan program on the federal budget, not the program itself. Whatever the intent, the action had profound implications for rural electric systems.

The Rural Development Act provided for insured loans at a 5 percent interest rate and guaranteed loans at a market rate. The insured loan, which had a term of 40 years, would be approved and administered by the federal government, then sold to an investor with the government insuring payment on the loan. The guaranteed loan would be made and serviced by a nongovernment lender. The government would guarantee the lender against loss. The shift to lending under the Rural Development Act would eliminate the 2 percent loan program.

It would also require systems to get approval for loans from local planning agencies and to make certifications regarding the unavailability of credit elsewhere under reasonable rates and terms. It was anticipated that there would be a charge for the government guarantees, which in some cases would be limited to 90 percent of the funds needed. As a further complication, there was an impoundment of REA funds that had been appropriated to be loaned and approved for lending. REA was unable to make loans between the effective date of January 1, 1973, and April 17, 1973, when the administrative procedures for operating under the new mandate were put into place.

Members of the rural electric program strongly objected to the Nixon administration's action at the end of 1972 that effectively terminated the REA program. CFC advanced funds to meet critical financial needs while rural electric leaders negotiated a new approach to financing.

"That really hurt, because there were cooperatives that had placed orders based on these funds being advanced," Paul Fox said. "I later heard many systems say that if it were not for CFC in December 1972, their system might not have survived."

CFC provided most of the program's financing requirements through its short-term loan program during the period that the REA loan program was suspended. "We did not make long-term loans to replace REA long-term loans," Smith said. "We advised the systems to obtain their capital from CFC on a short-term basis through the revolving line-of-credit program until something more suitable could be worked out. We did not want it to appear to Congress or to anybody else that CFC could or would replace the REA loan program."

Between January and April, CFC approved short-term loans totaling $44.6 million to meet the immediate capital requirements of its members. "It wasn't so much that we put many dollars out in that period of time. It was the fact that CFC was there, and we had the dollars for the systems," said Charles B. Gill, who was CFC's borrowers operations officer at the time.

"All of a sudden we had a situation in which CFC, which we had created to be a supplemental financing agency, of necessity had to become a primary financing agency. There wasn't any place else to go," said NRECA General Manager Bob Partridge.

Many of the member systems experienced an initial sense of

panic. Some systems were considering layoffs and other drastic cost-cutting actions. Smith and Gill visited cooperatives throughout the country to reassure them. "We told them not to do that, that this would pass and that if they needed money, we had it. We felt compelled to avoid a rush to defeatism and wanted to give the systems some sense of comfort that there was light at the end of the tunnel," Gill said.

Nixon Changes Unacceptable

The Nixon changes were unacceptable to the rural electric program. The action was the major focus of CFC's January 29–30, 1973, board meeting. "REA hopes to continue processing loans using the same criteria under the Rural Development Act as it had under the Rural Electrification Act," Administrator Hamil told the board. He also noted that several congressmen had expressed an interest in introducing legislation to amend the Rural Electrification Act to provide for an insured and guaranteed loan program and that some had expressed very favorable support for a program that would guarantee CFC loans.

Smith discussed with the board the position CFC should take on these issues, particularly if asked to testify before Congress. "We should recognize the impact of loans under the Rural Electrification Act on the federal budget and develop a financing program that eliminates this problem," he said. The CFC staff had been studying the effects of higher interest rates on borrowers and found that a tiered approach was needed for the distribution systems, with very low cost money for some systems, 5 percent money for others and a combination of 5 percent and market-based rates for the remainder. Another approach would be needed for the G&Ts. Given the fluidity of the situation, the board did not want to lock in a position at that time and instructed Smith to use his own discretion in responding to questions.

NRECA's first response was to support legislation to restore the 2 percent program under the Rural Electrification Act and to direct, rather than merely empower, the REA administrator to make loans. The bill, known as the Humphrey-Aiken bill, passed the Senate by a vote of 69 to 20. "Our position was that REA could not ignore their obligations to continue the traditional REA loan program," Smith said. However, it was clear in meetings with Office of Management and Budget Director Roy Ash, Agriculture Secretary Earl Butz and others that the administration intended to phase out all government financing for rural electrification as soon as possible. At a special board meeting on January 22, 1973, the NRECA board adopted a resolution strongly supporting the Humphrey-Aiken legislation but also saying that as soon as loan procedures under the Rural Electrification Act were reinstated, NRECA would negotiate in good faith with all parties concerned toward amending the Act to provide whatever changes might be required to improve its budgetary impact and/or interest rate structure if low interest rate capital would continue to be available to cooperatives that needed it.

The NRECA board appointed a Financial Advisory Committee, chaired by Louis Strong, who succeeded Smith as general manager of the Kentucky statewide association and was serving as president of NRECA, to make recommendations on a response to the situation. CFC President Vincent Slatt and board member Earl L. King served on the committee.

Timing was critical. The NRECA board preferred to solicit member input at the annual meeting before discussing legislative alternatives, but Congressman W. R. Poage, ready to introduce amendments to the Rural Electrification Act, said that would be too late. The Finance Advisory Committee met on February 6 and agreed on general principals for legislation, including amending the Act to provide direct insured and guaranteed loans; to create a revolving fund; and to vary interest rates in accordance with specified criteria.

At NRECA's 1973 annual meeting, the membership adopted the recommendation of Strong's committee that NRECA support the Humphrey-Aiken legislation and try to work out legislation creating an REA loan program with loans at 2 and 5 percent interest rates, a revolving fund and 100 percent government guarantees of private loans for rural electrification.

CFC started work on a second indenture for funding the guarantees. A triple-A rating, based on the government guarantee, was expected.

Administrator Hamil resumed approving long-term loans to distribution borrowers under the terms of the Rural Development Act on April 20, 1973, but most systems preferred to use short-term funding while the legislation for the new REA program was being considered. "Secretary Butz and Administrator Hamil did all they could to make the Nixon proposal work," Partridge said. "The Rural Development Act was a very poor mechanism and was not really intended to provide funding for rural electrification."

Legislation Creates New Loan Programs

Legislation amending the Rural Electrification Act of 1936 was enacted on May 11, 1973. It replaced the 2 percent direct loan program with a program of insured and guaranteed loans. It met the administration's goal of removing the annual appropriation for REA from the federal budget but kept the authority of the administrator within the provisions of the Rural Electrification Act, which were more workable than those of the Rural Development Act.

Under the revised program, insured loans were to be made at a standard 5 percent interest rate or a special 2 percent interest rate for systems with low consumer density or low average revenue per mile of line. The loans would come from the new Rural Electrification and Telephone Revolving Fund. Notes for the loans would be sold by the federal government to investors with REA insuring the purchaser against loss. The administrator was authorized to require supplemental financing from CFC or another credit source. Loans guaranteed by REA would be made by CFC or another lender at rates agreed on by the borrower and the lender. The government would guarantee the lender against loss of principal and interest.

CFC devoted substantial effort to the development of a program to fund the REA guarantees. It developed documents, engaged underwriters and, through NRECA, had legislation introduced in Congress to make technical changes in the Rural Electrification Act that would facilitate the issuance of guaranteed securities. The proposed rules for implementing the guarantee program published by REA specifically cited CFC as a qualified lender under the program.

The Federal Financing Bank (FFB) was established by legislation that went into effect December 29, 1973—exactly one year after Nixon precipitated the changes culminating in the guarantee program—for the purpose of providing coordinated and more efficient financing of federal and federally assisted borrowings. The Treasury Department announced on February 19, 1974, that the REA guarantee program would be eligible for financing through FFB. REA signed a loan commitment agreement with FFB on August 14, 1974. FFB offered member systems a more attractive way to fund the guarantees, and the CFC program was not implemented.

"What Mr. Nixon did for us in 1973 was the greatest thing he could have done and the worst thing he could have done," Gill said. "It created the guarantee program for the G&Ts, which made billions of dollars available that would not have been made available at the 2 percent rate. The good part was that it allowed us to have money for the G&Ts. The bad part of it was that it allowed us to make, in hindsight, some poor decisions. We wouldn't have been in some of the problem power plants if we had been relying on 2 percent money to do it."

In addition to establishing the insured and guaranteed loan programs, the 1973 legislation furnished a statutory base for supplemental financing by permitting the REA administrator, at his discretion, to provide a lien accommodation in any case in which he felt that the purposes of REA legislation could be achieved, rather than being limited by the stricter standards established by the General Accounting Office. CFC was asked to assist in drafting the legislation and did so.

The first five months of 1973 brought sweeping changes to the rural electric program. They also established CFC as an important part of the rural electric family, an organization that members could turn to in times of crisis and duress.

G&Ts Expand Throughout the 1970s

Jerome Katzin addressed a seminar for retail rate consultants sponsored by REA in May 1975. "In the past, Congress has from time to time restrained REA's resources and established limitations on how the monies will be used," he said. "Thus, for many years, REA had to restrict investment in G&Ts, so that the cooperatives were denied the ability to finance greater power self-sufficiency at a time when ample low-cost funds might have been used for that purpose. In retrospect, that has been a costly omission for rural America." In the mid-1970s, the rural electric program undertook a substantial expansion of generation and transmission resources to remedy that situation.

In the wake of the oil embargo, the co-ops' desire for independence, fostered in the Clapp years, became even more intense. Some traditional sources of power, including the federal power marketing agencies and investor-owned utilities, expressed doubt about their ability to finance needed generation additions, and the attitude of the power companies toward working jointly with

The G&Ts undertook a major capacity expansion program in the 1970s to meet member system needs that were growing faster than the national average.

cooperatives began to change. In 1972, CFC had 33 power supply members. By 1979, the number had increased to 50. In 1972, REA borrowers had 7,030 megawatts of generating capacity online, with an additional 1,061 megawatts under construction. By the end of 1979, 16,578 megawatts were in service, with an additional 13,487 megawatts of capacity under construction. During the mid-1970s, cooperative systems were investing approximately $9 in power production and transmission facilities for every $1

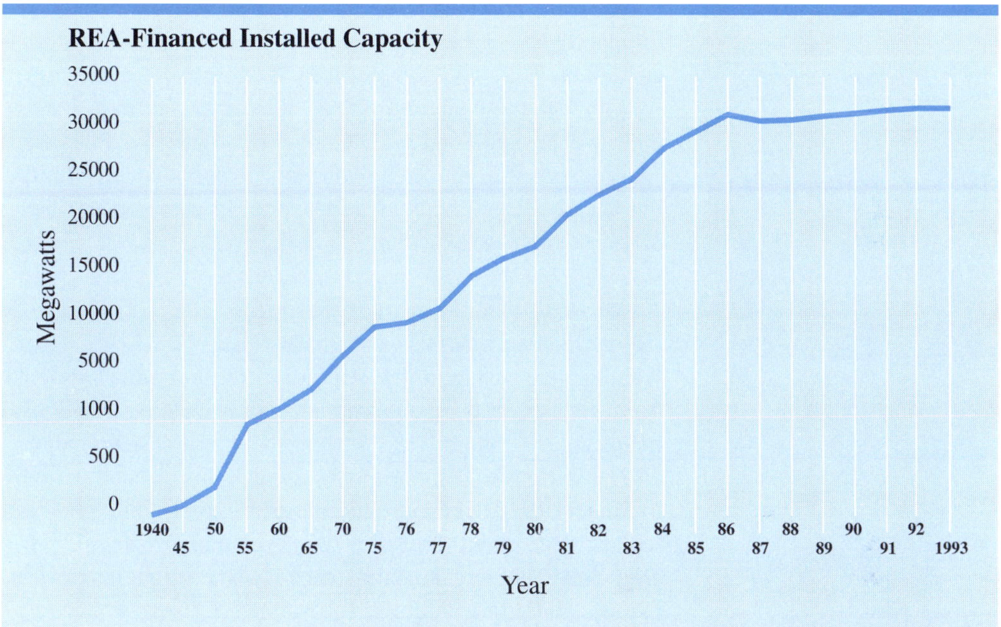

invested in distribution facilities. This could not have occurred at a worse time.

"The economics of the power industry give it an insatiable demand for additional capital of which only a portion can be generated from operations," Katzin said. "Rising prices, the inadequacy of depreciation to cover the cost of replacing like capacity and growth in market combine to force the utility to seek outside capital. For investor-owned companies, part of this new outside capital is equity; among cooperatives, all of it must be debt."

The utility industry was shifting from a decreasing cost structure, where each new unit was less expensive to build than the last, to an increasing cost structure, where each new unit was more expensive, sometimes very much more expensive, than the last. High inflation rates further increased costs. With the emergence of the environmental movement, systems were required to invest million of dollars in pollution control equipment, which added even more to the demand for capital. Construction lead times increased to a decade or more. In the high interest rate climate, this was a major factor in the increased cost of generating capacity. Many cooperatives became participants in nuclear plants, some of which were later canceled and all of which ultimately had higher capital costs than alternative forms of generation.

Although these factors affected all segments of the utility industry, the debt-financed G&Ts were particularly vulnerable. The cost pressures affected cooperatives more than other utilities because the capacity additions represented a greater share of their total resources.

CFC's Front-End Financing Program Facilitates Construction

Although the advent of the REA guarantee program eliminated the need for the 30–10–60 program and the bulk power supply financing program for G&T systems, CFC did develop many other programs to help the G&Ts through this difficult period. In January 1974, CFC introduced a new line-of-credit program, replacing the initial short-term loan program. The revised program provided more flexibility and allowed distribution members to borrow up to $1 million and power supply members up to $7 million for periods of up to 12 months on an unsecured basis.

In May 1974, CFC introduced an intermediate-term loan program to provide front-end financing for G&T projects during the period required for the G&T to establish feasibility and secure long-term financing through the REA guarantee program. The intermediate-term loans could be used for engineering and economic studies, site purchase, procurement of fuel supply and negotiating costs associated with joint participation projects. Intermediate-term loans were also used to help systems time drawdowns from FFB under the guarantee program.

Initially, the intermediate-term loans were unsecured. As the use of the intermediate-term loan program expanded and the size of the loan requests increased, CFC was able to negotiate security arrangements with REA.

"Once G&Ts got the REA guarantee program and the front-end financing program from CFC, they were in good shape in terms of access to funds," said CFC Engineering Officer Milton A. Chase.

CFC's front-end financing program provided a valuable source of funds for power supply systems during the most recent construction cycle.

The need for front-end financing was driven in part by changes in project requirements. "In the early days, any engineering firm would be happy to make a feasibility study for a G&T at a very low cost," Chase said. By the 1970s, that was no longer true. Most projects required detailed environmental studies, and the new joint participation projects with investor-owned and municipal utilities involved lengthy negotiations and extensive feasibility analysis. "CFC's short-term financing was very important in keeping the G&T project going while these things were being done," Chase said.

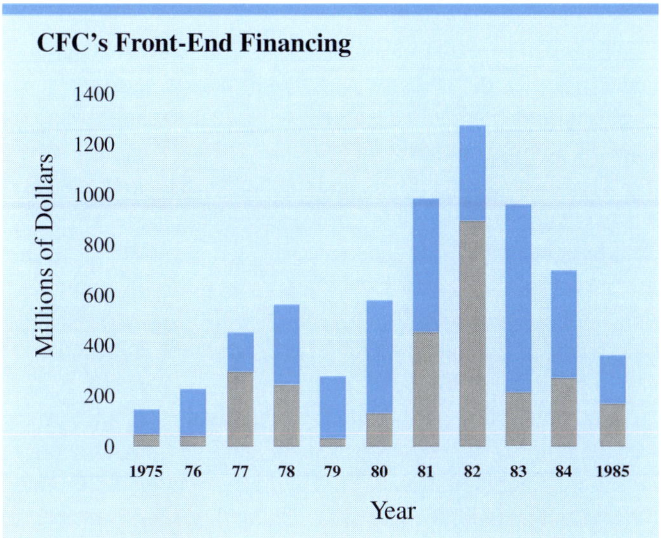

Lines of credit outstanding

Intermediate-term loans outstanding

CFC Guarantee Program Reduces Costs

The CFC guarantee program, introduced in 1977, allowed the G&T systems to substantially reduce the cost of pollution control facilities for their projects. "The program was suggested by the example of organizations such as the Municipal Bond Insurance Corporation, whose only business was to insure investors against defaults on bonds issued on a tax-exempt basis for municipal purposes, including pollution control," Fox said. "We had long discussions with Jerry Katzin and Russ Fraser [CFC's friend from Standard & Poor's who had moved to Paine, Webber] about how we could lend our good credit to our members' names so they could issue tax-exempt bonds at a favorable rate."

As much as 30 percent of the cost of a generating project was for pollution control facilities and could be financed in this way. CFC developed a program to guarantee and sell tax-exempt pollution control revenue bonds issued by a municipality or other governmental entity on behalf of a CFC member system. CFC was able to charge a very low fee for the service, because it was not required to earn a TIER on the guarantees. In most cases, CFC's guarantee gave the member system more credibility, a higher credit rating on the bonds than would otherwise be attainable and a lower interest rate. CFC also provided the member system with a package of expertise. CFC developed the indenture, dealt with the investment bankers, negotiated the price and helped to sell the issue.

The board appointed Kuhn, Loeb & Company as managing underwriter for the program with Paine, Webber, Jackson & Curtis as comanaging underwriter. Chapman & Cutler was selected as bond counsel.

"They were very complicated financings. The first one was a particularly difficult transaction," said Kuhn, Loeb's Robert R. Edmiston.

The initial offering of CFC guaranteed pollution control bonds totaling $109 million for nine member systems was sold in November 1977 at an effective interest rate of 6.28 percent. The comparable FFB rate at the time was 7.98 percent.

One of the things that made the initial issue so complicated was the number of cooperatives involved. The individual offerings were combined, because a larger issue would attract more interest and increase the marketability and liquidity of the bonds. Each cooperative had a different municipal issuer. "It was an enormous event because of the logistics of coordinating information on the cooperative issuers and the municipalities that would be sponsoring the tax-exempt transactions and then communicating all of that to the rating agencies and to the tax-exempt investors who were going to purchase the transaction," Edmiston said. "It was a very time-consuming process. The second transaction was much easier because we had put all of that in place."

The Art and Science of Pricing

The pricing for the transaction was further complicated because the cooperative facilities were located in different states. In some states, there were very few opportunities for investors to purchase tax-exempt securities. "Because of supply and demand you might get a better price for an Iowa piece of paper than you would for an issuer from Wisconsin, regardless of the credit rating," Edmiston said.

Part of Edmiston's job was to put the pricing recommendations in perspective for the issuers. The process was similar to the one for pricing CFC's collateral trust bonds. There were comparisons to the pollution control revenue bonds recently issued by investor-owned utilities and to the pricing on general obligation bonds issued by municipalities. There were comparisons between the general obligation bonds and revenue bonds recently priced. The analysis considered how well the other issues were received, what their credit ratings were and which state was the issuer. "All that came together to form a pricing judgment," Edmiston said. "By and large we got very good pricings."

There was no exact formula. Pricing a bond transaction is as much art as science. "You have to evaluate all the critical elements there and place them in perspective," he said. "You have to have the credit. You have to have the market. You have to have built the investor base. You must be sensitive to the needs of the investor base at various points in time. It does change. You are really trying in some respects to thread a needle."

The underwriter's job is to create demand for the transaction by emphasizing the strength of the credit and how it will perform over time. "You want to create as much of an oversubscription as possible so that you can get the very best price for the issuer," Edmiston said. "At the same time, you want to make sure that the transaction doesn't trade down in the aftermarket, because the investor does not want to feel that he bought the transaction at a high initial offering price when he could have bought it later in the secondary market at a lower price." It can be a delicate balancing act.

After the success of the initial long-term pollution control bond program, CFC developed other variations. In August 1978, the board approved a guarantee program to provide interim tax-exempt pollution control financing on a short-term basis while long-term financing was put into place. CFC arranged a master credit agreement with 14 commercial banks, effective December 1, 1978, to provide loans, guaranteed by CFC, for the interim period. In 1979, CFC expanded the short-term options by offering a guarantee of bond anticipation notes to provide interim financing.

Also in 1979, the board expanded the tax-exempt program to include guarantees of industrial development bonds and tax-exempt revenue bonds issued by public power districts.

"The guarantee program saved our members, primarily the G&T systems, millions of dollars," Gill said. "The Banks for Cooperatives wasn't allowed to guarantee tax-exempt bonds. We were the only one that could do that for our members."

Engineering Staff Provides Technical Support

Not all of the support needed by the G&T systems was financial. CFC had established an engineering department at the time it was preparing to issue Series A. Engineering Officer Chase and Senior Planning Engineer Sam Shiozawa helped to focus attention on the problems and opportunities facing power supply systems. The first Power Supply Symposium was held in Washington, D.C., on December 10 and 11, 1973. "CFC's purpose in sponsoring the symposium is to provide a forum for the meaningful exchange of information about a subject that is vital to the well-being of the rural electric program—the continued development of bulk power supply resources and facilities under changing financing conditions," Smith said at the time. The program included sessions on power pooling, fuel supply, environmental requirements, CFC financing, the public bond market and leasing.

"In retrospect, they were good meetings. The subject matter was interesting and topical," Chase said. "One of the things CFC did was to invite outside speakers to bring rural electrification out of its shell. If you knew only what was going on in the rural electrification power world, that was not sufficient—you couldn't understand what it was all about. You had to have a broader context."

For two decades the symposium was a CFC tradition that enabled G&T managers and their senior staff to exchange ideas and information with financial and technical experts and each other.

Fuel supply was another major issue for G&T systems during the 1970s. They placed a high priority on procuring reasonably priced fuel supplies for the large construction program then underway. Part of the nation's response to the oil embargo was increased reliance on coal technology. Many coal users moved to secure both supply and cost through long-term contracts.

Chase worked with the G&T Managers Association on a proposal for a cooperative fuel corporation to provide fuel reserves and supplies on a cooperative basis. Under the Chase concept, the organization would have an ongoing responsibility to deal with each G&T's coal supply and make recommendations, leaving the basic decisions with each G&T.

Although Chase's specific proposal was not adopted, two fuel supply cooperatives were organized in 1974. The Energy Resource Corporation attempted unsuccessfully to secure coal supplies by purchasing an operating coal company.

Western Fuels Association, Inc., was also incorporated in

CFC provided for a number of Western Fuels' projects, including locomotives and rail cars to transport fuel from mines to generating stations.

1974 as a not-for-profit fuel supply cooperative for consumer-owned electric utilities, including both cooperatives and municipals. Western Fuels both purchases and, through affiliates, mines coal for its member systems, with total coal deliveries of more than 21 million tons in 1993. The cooperative also owns and operates two rail lines and over 1,000 rail cars. CFC has provided financing for a number of Western Fuels' projects and facilities.

"The most significant thing we did on fuel supply was to sponsor a comprehensive study on coal supplies and reserves," Chase said. The study, conducted in 1976 by ICF, Inc., evaluated the resources available to every G&T cooperative. "This study gave the systems all the information they needed to develop fuel supplies. A company such as American Electric Power might not have found it useful, but many of our systems did not have experience in this area."

CFC also had discussions with REA about financing fuel supplies. REA informally agreed to provide a lien accommodation for any long-term fuel loans CFC might make.

The federal coal leasing program was a major topic at the 1977 Power Supply Symposium. Amendments to the Federal Coal Leasing Act gave preference to public bodies, including cooperatives, in obtaining federal coal leases. CFC worked with the G&T Managers Association to present 17 nominations for coal leases on reserves in North Dakota, Wyoming, Utah, New Mexico, Oklahoma and Alabama.

Power Supply Study Committee

The future cost and availability of power supply continued to be a concern throughout the 1970s. In May 1977, the CFC board authorized Governor Smith to work with NRECA in establishing a Power Supply Study Committee. The committee's charge was to make an in-depth study of various aspects of the rural electric power supply needs, including financing, joint ownership of facilities, types of generation and power supply, wholesale rate structures, fuel supply costs and availability, relationships between G&T systems and their members, REA requirements, load management and forecasting.

The committee followed the pattern established by the Long-Range Study Committee. "It was an interesting example of the way in which the cooperatives in general and CFC and NRECA in particular tackled the local problems that covered the whole country," Chase said. "They set up a committee that had a proper balance by regions and a proper balance by managerial people and board people and sometimes consultants. Then there would be regional meetings at different places to get input. That is one of the ways in which the rural electric people kept themselves together. It is a very good system because it gives everybody an opportunity to have their say. It is a very bad system in that you can't do all the things everybody wants."

The Power Supply Study Committee was chaired by NRECA board member Louis Strong, president of the Kentucky statewide association. The vice chairman was CFC board member Donald R. Norris, manager of Eastern Iowa Light and Power

The Power Supply Study Committee made an in-depth study of various aspects of rural electric power supply needs.

Louis Strong
Chairman
President, Kentucky Association of Electric Cooperatives
Louisville, Kentucky

Donald R. Norris
Vice Chairman
Manager, Eastern Iowa Light and Power Cooperative
Wilton, Iowa

Stanley K. Bazant
Executive Vice President and General Manager, Plains Electric G&T Cooperative, Inc.
Albuquerque, New Mexico

L. P. Beverage
Manager, Four County Electric Membership Corporation
Burgaw, North Carolina

Gerald F. Diddle
General Manager, Associated Electric Cooperative, Inc.
Springfield, Missouri

Kenneth Gordon
President, Newton Country Rural Electric Membership Corporation
Kentland, Indiana
President, Wabash Valley Power Association
Indianapolis, Indiana
Director, Indiana Statewide Rural Electric Cooperative, Inc.
Indianapolis, Indiana

Leo E. Grubl
Director, Rushmore Electric Power Cooperative, Inc.
Rapid City, South Dakota
Director, West River Electric Association, Inc.
Wall, South Dakota

James M. Hubbard
Executive Vice President, North Carolina Association of Electric Cooperatives
Raleigh, North Carolina

Lloyd M. Lange
President, Badger Power Cooperative
Wautooma, Wisconsin
Director, Dairyland Power Cooperative
La Crosse, Wisconsin
Director, Columbus Rural Electric Cooperative
Columbus, Wisconsin

Philip O. Martin
General Manager, United Power Association
Elk River, Minnesota

William T. Nordeen
Manager, Northern Lights, Inc.
Sandpoint, Idaho

Dick Surrency
Director, Clay Electric Cooperative, Inc.
Keystone Heights, Florida
Alternate Director, Seminole Electric Cooperative, Inc.
Tampa, Florida

George B. Taylor
General Manager, South Mississippi Electric Power Association
Hattiesburg, Mississippi

C. R. Thiessen
President, Basin Electric Power Cooperative, Inc.
Bismarck, North Dakota

Reuben Yoselson
Director, Allegheny Electric Cooperative
Harrisburg, Pennsylvania
Director, Claverack Rural Electric Cooperative
Towanda, Pennsylvania

Cooperative. The remainder of the committee included five members from the boards of directors of rural electric cooperatives, five distribution system managers and five G&T managers. The committee held its first meeting in December 1977 and established three subcommittees to address specific areas: power supply adequacy and reliability; the financial and economic impact of power supply; and the government, industry, member and public relations aspects of power supply. Subsequently, the committee appointed numerous special task forces, and 16 task force reports were issued in addition to the committee's final report. Six regional hearings were held in April 1978.

The committee's report, Electric Power Supply... Issue of the 80s, was presented to the membership at the 1979 annual meeting. More than 10 years after the Rural Electric Financing Study Report and the work of the Long-Range Study Committee, cooperatives were still experiencing higher levels of growth than other utilities, with a 6 percent rate of growth projected for the next decade. The report recognized the need to reduce reliance on imported oil as a major source of energy through conservation and development of coal and nuclear resources and to restructure the regulatory and environmental requirements that impeded doing that.

The committee's 13 recommendations addressed energy conservation and development issues, regulatory reforms and cost reduction, and consumer education and action programs.

"It was a broad program with no specific issue that captured the imagination to make it memorable," Chase said. Despite that, many of the committee's recommendations were implemented. The Technical Advisory Committee established task forces on nuclear energy, research and development and regulatory reform. More cooperatives became involved in industry research through the Electric Power Research Institute. As a result of the committee's effort, the rural electric program became more involved in conservation, nuclear oversight, energy research, technology development and consumer education.

Member Services Programs Strengthen Members

When Charles Gill joined CFC as borrowers operations officer on April 1, 1972, he was attracted by the challenge of something new and exciting. "I thought this would be the organization of the future," he said. "My job was, number one, to make sure the membership understood what the concept of CFC was and the role it was going to play with them; second, to develop a rapport with the membership so they were supportive of CFC; and last, to develop a technical capability to assist the members in being more efficient and effective."

With Smith's encouragement, Gill generated an explosion of new member service programs. Undoubtedly the most important one, the one that changed rural electric operations the most, was the cash management program. In 1972, rural electric systems did not have sophisticated programs for managing their money. "It was politically unacceptable to make money on 2 percent funds borrowed from REA," Gill said. Systems usually kept any excess in a checking account or a savings account paying 2–3 percent interest. The feeling was that it didn't cost the co-op very much, and it avoided criticism of the link with the 2 percent program.

With the changes in the REA program in 1973, the cost of funds was much higher to most systems, and economic conditions precipitated a dramatic increase in interest rates and inflation. "Banks were making millions of dollars off of the cooperatives by keeping their money, investing it elsewhere and providing the cooperative no real return on it," Gill said.

Commercial Paper—Lower Cost for CFC, Higher Returns for Members

One of CFC's primary goals was to borrow funds for its members at the lowest possible cost. In 1973, a creditworthy company could borrow funds more cheaply by issuing commercial paper than it could by getting a loan from a bank. Commercial paper is a promissory note that normally extends no longer than 270 days and may be for as short a period as one day. A company's commercial paper was usually sold through an agent, such as an investment bank. The CFC board approved development of a commercial paper program during its November 1973 meeting. The program was implemented in May 1974, when CFC began issuing commercial paper to other corporations through Goldman, Sachs & Company. (At the time, Kuhn, Loeb did not act as an agent for commercial paper sales. Goldman, Sachs was the largest firm in the paper market.) CFC's commercial paper was backed by bank lines of credit and received the highest possible ratings, Prime-1 from Moody's and A-1 from Standard & Poor's.

Once the program was established and operating well, Smith had the idea that made Gill's cash management program successful—selling commercial paper to the CFC member systems. "We were offering our members an instrument, commercial paper, that they could take advantage of with the very low minimum investment of $5,000. If they were to invest in commercial paper in the money market, they would have to have about $500,000 to $1 million to invest," Fox said. "The ability to invest small amounts of money at market rates was very attractive to our members." CFC also instituted the sale of CFC notes with maturities of 30, 60 and 90 days to enable member systems to invest relatively small amounts of money for longer periods at attractive rates. Although the commercial paper program introduced CFC members to a whole new concept of cash management operations, it also became an important source of attractively priced short-term variable-rate capital for CFC.

CFC's senior staff concentrated on the dual goals of obtaining adequate amounts of funding at attractive rates and helping its member systems improve and maintain financial performance. Pictured, **left to right:** ***Angelo Salera, controller; Ira Shesser, loan officer; John Jay List, corporate counsel; Charles B. Gill, deputy governor; J. K. Smith, governor; Nicholas J. Walsh, administrative officer; Clyde Denton, information and member relations officer; Gary A. Gordy, borrowers operations officer; Milton A. Chase, engineering officer; and Paul Fox, finance officer.***

CFC opened the Money Desk in September 1974 to administer the member commercial paper and CFC note programs and to manage CFC's own cash investment activities. To take advantage of the program, a member system simply called the Money Desk to arrange its investment, which was made and returned through bank wire transfers between the member system's bank and CFC's depository bank.

"We couldn't have done this in the private market," Fox said. "We couldn't have gotten Goldman, Sach's list of investors. By selling to our own members, the operation was already in-house. We had the lists. We knew the people. We had access to them."

Starting in 1975, Gill offered a series of cash management workshops to teach the member systems how to benefit from the new program. "We talked about the time value of money, because our cooperatives never had to worry about that before," he said. "We went through the mechanics of how you move money. We showed them what banks might make on interest-free

CFC commercial paper provided an important source of capital for CFC and an attractive investment opportunity for member and nonmember systems.

money. We showed them how to develop a cash flow analysis to determine what their cash needs and resources were. Then we coupled it with our line-of-credit program to show them how, if they misjudged, they could use the line of credit to fill the gaps of cash needs."

More Revenues, Same Rates

The response to the program was excellent. "They saw for the first time how they could increase their revenue without increasing rates to their members," Gill said. "In dollars and cents, this is one of the bigger things we have done for the membership. They weren't large enough to enter the money markets, and the money markets didn't know about them. Our commercial paper program gave them an avenue they didn't have before to invest their capital outside local channels."

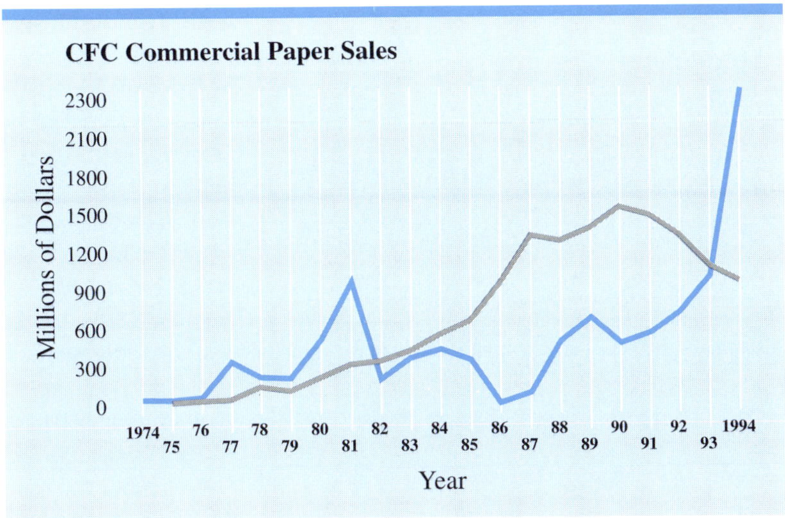

Dealer

Member and nonmember

The local banks were sometimes critical of CFC for taking those assets out of the community. CFC responded that the members were free to leave their investments in the local community, but that they should share in the benefits of earnings on their funds. "We gave them leverage to develop a greater interest income from their local banks," Gill said.

The CFC staff conducted 56 credit and cash management workshops in 1975. As the systems implemented what they learned, nonoperating margins (which include interest income) for distribution systems almost doubled in a three-year period, from $33.2 million in 1976 to $62.6 million in 1979.

The member commercial paper–cash management program also demonstrated the value of CFC's self-help concept and the diversity of the rural electric program. "If there is an earthquake in Oregon and they need money fast, the systems in Florida can lend it to them by investing in commercial paper," Fox said. "The Florida systems that are investing are making money, the Oregon systems are paying least cost money for their short-term capital and getting it quickly, and it all comes through their own intermediary in Washington. It's a wonderful story."

The program also drew a positive response from private investors. "This was one aspect of our program that warmed the hearts of investors," Fox said. He cited member commercial paper investments along with CTC investments and retained patronage capital as an important part of the total member contributions to CFC funding. "It was a tremendous selling point for CFC to have member investments of that magnitude."

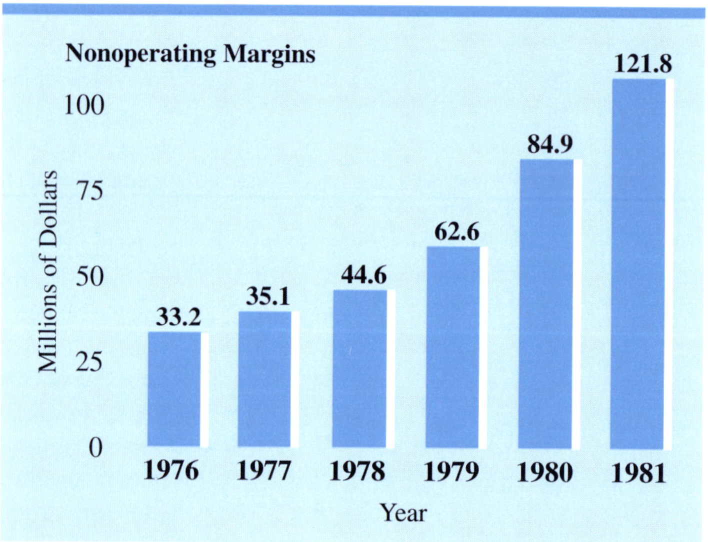

Nonoperating margins for distribution systems increased dramatically after CFC initiated its cash management workshops.

In some respects, CFC's cash management program became a victim of its success. "Our membership has become more sophisticated about their alternatives, and now they are investing in a lot of other financial instruments that give better returns than our commercial paper," Gill said. "We can't give better returns because we can borrow in the market as cheaply as we can issue member commercial paper. We can't give returns that raise the rates for the rest of our members. But we still provide the yardstick for the Street and other investment opportunities."

Series E Bonds

The success of the member commercial paper program encouraged CFC to develop other opportunities for members to invest in CFC. In November 1975, the board authorized issuance of a new series of collateral trust bonds for direct sale to individuals in the rural electric community. The $25 million issue was designated Series E and sold through member system representatives.

CFC member systems, their consumer members, and present and former systems directors and employees could invest in the bonds of two- to five-year maturity. The bonds were offered in $1,000 denominations.

In May 1978, the board authorized issuance of $50 million in Series E-1 collateral trust bonds. The Series E-1 bonds were similar to the Series E bonds and offered additional maturities of six and seven years. In May 1980, the board authorized Series E-2, which carried a floating interest rate.

The Series E program's greatest value was in giving members of the rural electric family an opportunity to participate directly in CFC. "It's a chance to invest in your own organization," Smith told members.

Although by no means a failure, neither was this program a major source of capital for CFC. The maximum aggregate amount invested was $23.6 million in 1979.

"When people invest, they invest to make money," Fox said. CFC's offerings had to compete with many other financial instruments, some of which offered more attractive rates and more flexible terms to individual consumers. The program was discontinued effective April 30, 1985.

Fox said CFC learned a lesson from the experience. "If we are going to do something creative, let's do it in an area where there are no alternative opportunities that are equal or better," he said. CFC was most successful when it offered programs to members, such as the cash management program or the pollution control bond guarantees, that enabled them do something they would otherwise be unable to do.

Capital Credits Refunds Demonstrate Value of Cooperatives

Capital credits, also referred to as patronage capital and member equity, are an electric cooperative consumer's payment in excess of the cost of providing electric service. It is the primary source of equity for an electric cooperative. Cooperative bylaws generally view each dollar of patronage capital as an investment in the organization that must be returned to the patron at some time, but the REA mortgage required a system to obtain a 40 percent equity level before making a patronage capital distribution. In 1973, only 93 rural electric systems were making general patronage capital retirements, or refunds. CFC wanted to change that.

"The capital credits concept was introduced early in the program to help justify the nonprofit characteristics of the rural electric systems," Shesser said. The IRS granted tax-exemption on the basis that the cooperatives would be nonprofit organizations and return the capital credits to their members. "Unfortunately, we didn't have a significant pattern of actually revolving capital credits in the rural electric program. Many of the systems were not even doing an effective job of keeping records, so that if they wanted to revolve capital credits they would have difficulty in doing so."

In November 1973, the board authorized the appointment of a committee to study and make recommendations on capital credits issues, subject to receiving the support of NRECA and REA to address the subject. Earl King, executive vice president of the Iowa Association of Electric Cooperatives, agreed to chair the Capital Credits Study Committee. NRECA subsequently agreed to cosponsor the group.

Not all of the rural electric systems were enthusiastic about the study. "Some people said that it was zero cost capital and that you shouldn't give back zero cost capital, only to borrow at 7 to 10 percent," Gill said. "The Capital Credits Study Committee was an attempt to drive home the fact that we have to return capital to the membership. We just can't keep it indefinitely."

The Capital Credits Study Committee followed the pattern established by the Long-Range Study Committee. The committee members, some of which were appointed by CFC and some by NRECA, represented a cross section of the rural electric program. The committee was open in its deliberations and solicited comments from anyone who had something to offer.

"Even though CFC has taken the initiative, it has never intended this study to be exclusively a CFC study," King told participants in the 1974 annual meeting. "We must listen to every suggestion and idea. We must analyze every approach. We must be completely objective. But even more important, this study must be a cooperative effort and include all segments of the rural electrification program." To be sure that everyone had an opportunity to contribute, the committee held a series of open hearings the week of June 10, 1974, that were attended by over 1,000 people from 400 cooperatives.

The committee's final report and recommendations were presented during the NRECA and CFC annual meetings in

February 1976. It supported the principal that each cooperative system should return patronage capital within the framework of building a financially sound organization. "The big thing that it did was to develop an equity management program that gave the cooperatives a better concept of how to set rates relative to margin levels and rotating capital," Gill said. "We provided a road map. They could still make individual decisions and have different levels. We provided a framework that said, 'If you do this,

The Capital Credits Study Committee focused attention on the value of rotating patronage capital.

Earl L. King
Chairman
Manager, Iowa Association of Electric Cooperatives
Des Moines, Iowa

Richard Arnold
Manager, Union Rural Electric Association, Inc.
Brighton, Colorado

C. D. Branch
Director, Brunswick Electric Membership Corporation
Shallotte, North Carolina

Richard Foster
Manager, Atchison-Holt Electric Cooperative
Rock Port, Missouri

Jack Goodman
Manager, Central Kansas Electric Cooperative, Inc.
Great Bend, Kansas

Walter Harrison
Manager, Georgia Electric Membership Corporation
Millen, Georgia

John Hill
Manager, Covington Electric Cooperative
Andalusia, Alabama

Harold Johns
Director, Polk Burnett Electric Cooperative
Centura, Wisconsin

J. L. Johns
Manager, Upshur Rural Electric Cooperative Corporation
Gilmer, Texas

D. L. Knight
Manager, Bossier Rural Electric Membership Corporation
Bossier City, Louisiana

Leland Leatherman
Attorney, Arkansas Electric Cooperative, Inc.
Little Rock, Arkansas

Philip O. Martin
Manager, United Power Association
Elk River, Minnesota

Lyle McCormick
Manager, Guernsey-Muskingum Electric Cooperative, Inc.
New Concord, Ohio

Paul Ogier
Director, Midwest Electric Membership Corporation
Grant, Nebraska

LeRoy Schultz
Manager, Chugach Electric Association, Inc.
Anchorage, Alaska

Dexter Smith
Manager, Arizona Electric Power Cooperative, Inc.
Benson, Arizona

Douglas Smith
Manager, Somerset Rural Electric Cooperative, Inc.
Somerset, Pennsylvania

H. L. Spurlock
Manager, East Kentucky Rural Electric Cooperative Corporation
Winchester, Kentucky

Leon Weaver
Manager, Talquin Electric Cooperative
Quincy, Florida

Roger Westenbroek
Manager, Top O'Michigan Rural Electric Company
Boyne City, Michigan

Fred Whittle
Director, Mountrail-Williams Electric Cooperative
Williston, North Dakota

George Wilcox
Director, Douglas Electric Cooperative
Roseburg, Oregon

this is what is going to happen for various growth rates, rotation cycles, costs of money and so forth.' "

Capital Credits Workshops Illustrate Options

As a follow-up to the committee's report and recommendations, CFC offered a series of two-day capital credits workshops to provide systems with guidelines for developing financial objectives and a plan for meeting them. The workshops included a hands-on session where participants could use the methods developed by the committee to analyze options for their own systems.

CFC also provided leadership to the program by developing and implementing its own capital credits program. In February 1977, the board approved an equity management plan based on a seven-year rotation cycle using the "first-in, first-out" (FIFO) method prescribed by CFC's bylaws. Under the FIFO method, patronage capital retirements, or paybacks, are based on the length of time the funds have been invested in the cooperative, with the oldest capital being returned first. (At the 1992 annual meeting, CFC's membership approved a bylaw change that allowed the board of directors to determine the method of retiring patronage capital. The board subsequently adopted the percentage method. Under the percentage method, patronage capital retirements are based on a percentage of the total capital credits on the books of a cooperative.)

CFC made its first refund of capital credits on February 1, 1980. It mailed checks totaling $2.6 million to member systems for patronage capital allocated during fiscal year 1973. The amount represented almost half of the interest systems paid on loans in 1973.

CFC's efforts in this area were successful to the degree that, by 1994, about 500 systems were retiring capital credits on a regular basis. "What we didn't sell, and the key part of it in my mind, was a different kind of retirement other than first-in, first-out," Gill said. Most bylaw provisions governing patronage capital rotation, including CFC's original bylaws, were based on an REA model from the early days of the program. "The patronage capital system that was created early on in the program was based on a static population, so it didn't matter which method you used." By the 1980s, population patterns had changed, and the FIFO method meant that a member who finally became eligible for a refund was likely to be living in another area purchasing power from another supplier.

"You might as well burn the money in terms of what it does for your cooperative other than meeting your legal obligation of disbursing the assets back to the member at some time," Gill said. "By retiring capital credits on a percentage basis, we felt that the current consumers would see a rebate now, an ownership interest in the system, and better understand the philosophy of cooperatives. If they had to wait 20 or 30 years to get their patronage capital, they wouldn't have the same feeling."

CFC Programs Support Management

There were many other innovations in member services programs during Gill's tenure as borrowers operations officer. For example, CFC first offered the REA check delivery service in January 1975. CFC would deliver a system's debt service payment check to REA on the exact date the payment was due. "On average, that program provided member systems with 60 days income interest a year," Gill said.

The program was expanded to a paying agent service in 1977. Under the enhanced service, a member could invest payments due to CFC, REA, wholesale power suppliers or others in CFC commercial paper maturing on the due date. CFC then wired the proceeds of the commercial paper investment in payment to the designated recipient on the day it was due.

CFC developed the Key Ratio Trend Analysis (KRTA) in 1975 as a way of analyzing the performance of distribution members. The KRTA calculates 52 ratios related to financial performance, expenses, power cost, growth and other key areas for a system for a five-year period. The analysis compares each ratio for the individual system to the median of all systems in the data base, which includes all REA and CFC borrowers, and ranks its performance on that ratio with all systems. The KRTA also provides comparisons based on the composite of all systems in the same state, with the same primary power supplier, of approximately the same consumer size, and with a similar rate of growth in total utility plant.

The analysis was originally developed for use by the loan department in its credit analysis and to help CFC identify systems that needed special assistance. The loans CFC pledged as collateral for its collateral trust bonds were required by CFC's indenture to meet certain quality standards. A distribution borrower was required to maintain a minimum TIER of 1.50 and a minimum DSC of 1.25 for its mortgage to be eligible collateral. Power supply systems had a lower standard. If a member system whose mortgage had been pledged failed to meet the required standard, CFC had to substitute good collateral or the authorized level of financing would be reduced.

"Obviously, the better the quality of the portfolio that secures CFC's bonds, the more attractive its bonds become. This produces a lower interest cost, which CFC can then reflect in reduced interest rates on its loans to its member borrowers," Katzin said. The KRTA helped CFC identify systems not meeting the indenture standards and those experiencing a downward trend in financial performance. CFC could then use the analysis to help the system identify specific problem areas and find ways to correct them.

Katie McCall processes a CFC paying agent service transaction. The paying agent service allowed member systems to invest payments due to CFC, REA or others in commercial paper until the due date when CFC would wire the commercial paper proceeds to the designated recipient.

The KRTA Is Useful Management Tool

CFC soon realized that the KRTA could be a useful tool for all systems themselves to use in analyzing their operations. CFC began providing a copy to each member as soon as it was available. "The KRTA turned into a management tool for the systems instead of just a credit analysis tool for CFC. It was very effective," Gill said.

Conservation was an important issue for consumer-members during the 1970s. CFC introduced a weatherization loan program to support conservation goals in May 1977. The program offered loan funds to member distribution systems that they could then relend to their members for installing insulation, storm doors and windows and other energy-conserving equipment in their homes and places of business.

Regulated systems turned to CFC for support with their local commissions. CFC began to educate the regulators about the concept of CFC and to encourage the commissions to consider an approach to ratemaking that would allow the member to go to the market through the facility of CFC.

The first service coordinators (later called area representatives, then regional vice presidents) joined CFC in 1974 to support these efforts. "The concept of the service coordinator originally was to put someone out in the field to better communicate with the membership and to get better feedback from the membership as to what the membership wanted and needed," Gill said.

The number of service coordinators was small relative to the number of member systems. "Our theory was that we ought to be there when we needed to do something specific with the member, but we shouldn't have so many field reps that it became a routine, coffee-drinking kind of job," Gill said.

In addition, because the members owned CFC and its cost of money was higher than that of REA, Gill knew more would be expected of CFC field representatives. For that reason, experience as a rural electric cooperative manager was a basic requirement for the first service coordinators. "We needed somebody who could relate to what was happening at the co-op, who had a feeling for the problems and issues," Gill said.

Originally, the service coordinators functioned as management consultants. They provided technical assistance on issues such as mergers, equity management and capital credits studies. They taught systems how to develop cash and equity management programs and interpreted CFC's policies. They also encouraged support for CFC.

"It was a very positive relationship," Gill said. "Everything we did grew the cooperative greater margins, grew their mentality, grew their training and grew their people."

CFC's role in providing that kind of support changed as the needs of the member systems changed. Over time, the relationship shifted to more of a partnership. "It has changed from one of feeding people to one of teaching them how to plant to the third phase of trying to convince them that we should be working with them to reap the harvest," Gill said.

In the 1990s, the regional vice presidents focused on marketing the various financial products available and helping member systems determine which were appropriate to their needs.

CFC Achieves AA Rating

CFC maintained a strong presence in the financial markets throughout the 1970s. The Series A collateral trust bond issue was followed by Series B in 1974, Series C and D in 1975 and Series F and G in 1979. The amount of debt raised by these issues totaled $550 million.

Standard & Poor's upgraded the bond rating on CFC's collateral trust bonds to AA with the issuance of Series D. "In view of the long proven track record of the REA electric cooperatives, the protection afforded the CFC bondholder, the geographical diversity of the member cooperatives and the professional management which CFC has exhibited since its inception in 1969, we have rated this issue of bonds AA and have raised the outstanding rating on the other series of bonds from A+ to AA," said Standard & Poor's in its publication *Fixed Income Investor*.

Moody's upgraded its rating of CFC collateral trust bonds to Aa from A with the issuance of Series G.

"This AA rating signifies many things," Katzin said shortly after the upgrades. "It shows that CFC was soundly conceived by its founders and that its organization and purpose have stood the market test of time. It also shows that the market has gained confidence from experience with the management of CFC, that it has demonstrated that it is being prudently and responsibly operated and managed. Finally, it shows confidence in the sustained quality of CFC's member borrowers, for after all, the fundamental credit of CFC rests in the final analysis on the quality of its loan portfolio."

One of CFC's challenges was to maintain its level of financial performance needed to support the high credit ratings. "Al Esokait of Moody's told us that we would have to maintain a 1.10 TIER if we wanted respectability in the financial community. Every time I met with Standard & Poor's or Fitch, they would say the same thing. 'You're good, but you've got to demonstrate your goodness,' " Fox said.

Earning a 1.10 TIER often meant earning margins that were, in the eyes of J.K. Smith, excessive for a cooperative. There were many discussions about the appropriate standards. Smith always had the final word with, "Paul, your problem is, you don't understand cooperatives."

The line became a standard response around the CFC offices. "Whenever we had an argument with Paul, we would say, 'Well, Paul, you just don't understand co-ops,' " Ira Shesser said.

Fox, of course, understood cooperatives very well and used the knowledge to gain access to new sources of capital for CFC. On February 26, 1973, CFC entered into its first five-year revolving credit agreement. The $100 million loan facility with 23 major banks was established to provide a source of funds for CFC's front-end financing program for G&T systems.

Once its domestic financing program was well established, CFC approached the international markets. "It became quite obvious to most of us in the financial world in the 1970s that this was a global economy," Fox said. "Getting into the international market was the natural, intelligent thing to do, because, among other things, there was a lot of money out there. There was a lot of financial expertise, and the banks wanted to become lenders to us."

CFC maintained a strong presence in the bond market throughout the 1970s.

Series	Issue Date	Due Date	Amount (in millions)	Interest Rate (%)	Credit Rating Moody's	S&P	Fitch
A	December 1972	2007	$ 50	7.4	A	A	Not rated
B	October 1974	1981	100	10.5	A	A+	Not rated
C	January 1975	1985	100	8.95	A	A+	Not rated
D	October 1975	1985	100	9.125	A	A+	AA
F	March 1979	2009	100	9.75	A	AA	AA
G	August 1979	2009	100	9.75	Aa	AA	AA

CFC often benefitted from the competition among foreign banks. Banks located in the same country frequently reviewed CFC's bank line list to compare their position relative to their competitors. If they considered themselves to be the lead bank in a particular country, then they wanted to have the top position in CFC's credit. Consequently, some banks increased the amount of credit offered to CFC to maintain their position.

The Move to CFC Square and Other Transitions

Although not directly related to loan or financing programs, two significant events that occurred during the 1970s demonstrated that CFC was now a well-established, viable organization. First, CFC purchased property for a headquarters facility in the Georgetown area of Washington, D.C., in May 1975. Then, there was an orderly transition of power as the last of the original CFC board members completed their terms of service in 1977 and Charles B. Gill succeeded J. K. Smith as governor in 1979.

The progress of CFC in establishing itself as an organization was reflected by its progression through a series of office spaces. When CFC officially opened for business on September 15, 1970, it made its first move out of the temporary space provided by NRECA into its own offices in the Blue Cross and Blue Shield building at 500 12th Street, S.W. The move was significant because it demonstrated CFC's intent to operate as an independent organization. CFC outgrew that location in just a few months and tripled its space by moving into the Reporters Building (court reporters, not newspaper reporters) at 300 7th Street, S.W., on June 1, 1971. It was clear that this was a temporary fix, and CFC initiated discussions with NRECA, National Telephone Cooperative Association and the Farm Credit Association about the possibility of jointly owning a cooperative organization building. That was an idea whose time had not yet come.

By February 1972, CFC had 34 employees. Administrative Officer Nicholas J. Walsh was able to lease more space in the Reporters Building, but the building owners made it clear that leasing any more space in the building would be difficult.

"Every time we moved, we thought we had moved into a space that would be big enough to meet our needs, and every time we guessed wrong," CFC President Vincent Slatt said. There was a consensus that CFC should own its own building. The argument centered on where and what type of building.

Walsh first approached the owners of the Reporters Building about buying the building. "It was owned by a partnership, and one of the partners said absolutely not," he said.

CFC then arranged to purchase the Army Times building in southwest Washington. Its owners were building a new facility in Springfield, Virginia. "The chairman of their board had a premonition the night before that they shouldn't do it," Walsh said. "When our attorneys showed up with the check and the contract, he talked his board into scrapping the deal."

Then CFC found some property vacated by Washington Gas Light Company on the industrial warehouse side of M Street in Georgetown. There were two buildings, one of about 25,000 square feet and another of about 100,000 square feet. The buildings could be purchased or leased. In January 1975, the board authorized the staff to lease the smaller building and to continue negotiations for purchase of both, subject to board approval.

CFC experienced an orderly transition of leadership when Charles B. Gill* (left) *succeeded J. K. Smith* (right) *as governor on September 1, 1979.

"The property had a lot of potential," Walsh said. "There were plans in discussion for giving a facelift to the south side of M Street, but there was not too much action yet, so it was a very attractive purchase price."

However, the board had not seen it. Smith arranged a site visit for the March 1975 board meeting. "The timing wasn't good, because the company that had vacated the structure had taken from it everything they possible could," Slatt said. "Some of the floor tiles were missing. Where they had disconnected the telephones, telephone wires were sticking out of the walls. You could see where they had removed some of the lighting fixtures. All in all, the facility didn't make a very good impression."

On the way back Smith told Slatt not to bring the purchase issue up in the board meeting. "I don't think the board will vote for it," he said.

Slatt figured the issue would have to come up, because the board had taken time to see the building. He made it the first item on the next morning's agenda. "We became embroiled in a very lengthy discussion," he said. "The opposition was not to spending the money to own a building but rather to the design and location of the building. Some of the board members couldn't see buying this run-down, ramshackle old building and then spending millions of dollars to refurbish it. It was their opinion that even then we would have something that would be totally unsuitable for one of the rising financial organizations in the United States."

Others on the board disagreed. They looked at the potential of the Georgetown location and thought it could be a nice place. On a vote of 12 to 10 the board authorized the governor to purchase the property for $3.25 million. The property was purchased from Washington Gas Light Company in April 1975.

It turned out to be a very sound investment. Part of the credit goes to the location and part to the renovation. "One of the wisest things we did was to hire Arthur Cotton Moore as the architect," Slatt said. "He renovated the North Building by putting in an atrium and overnight the building really looked presentable."

Renovations on the North Building, the smaller space, were completed in October 1975 at a cost of $565,000. (A common visitor's question at the time was, "Who is Mr. North?" North, of course, signified its location on the property.) Renovations on the warehouse facility took longer and cost more—$5 million, including a parking garage and plaza. It was dedicated as the J. K. Smith Building on December 12, 1978.

CFC sold the facility at the top of the market in 1989 for $21.5 million when a move to new facilities in Herndon, Virginia, marked yet another stage in CFC's evolution. "We did well on it," Walsh said.

Seamless Transition in Board and Management

The Long-Range Study Committee indicated its view that the organization CFC transcended individuals when it limited directors to two three-year terms. The first board elected by the membership held its first meeting in March 1971. One of the

About CFC Square

When construction began on CFC Square in April 1975, the Georgetown section of Washington, D. C., where it was located was at the beginning of a period of revitalization and economic expansion. A unique blend of homes, townhouses, shops, boutiques and restaurants, it was soon a prime location for office space, too.

The CFC Square complex included two office buildings, the J. K. Smith Building, named in honor of CFC's first governor, and the North Building, named for its location on the site. The two buildings provided CFC with a total of 125,000 square feet of office space. The Smith Building was adjacent to the scenic Chesapeake and Ohio Canal and overlooked the Potomac River and the John F. Kennedy Center for the Performing Arts.

In addition to the two office buildings, the site included a residential space. Georgetown Mews Associates, a Washington development firm, leased airspace over a section of CFC Square and constructed a 139-unit apartment building, which it owned and managed. The residential facility helped CFC achieve maximum utilization of the property under local land-use and zoning regulations.

CFC acquired the 57,000 square foot site from Washington Gas Light Company in April 1975 at a cost of $3,125,000. What became the North Building was an office building, but the soon-to-be Smith Building was a warehouse facility. CFC retained the architectural firm of Arthur Cotton Moore/Associates to design the new complex.

Renovation of the North Building, which was originally constructed in 1957, began in May 1975 and was completed in October 1975 at a cost of $545,000. It gave CFC approximately 25,000 square feet of office space on five floors.

Renovation of the J. K. Smith Building began in November 1976. Most of the existing structure, which was originally constructed in 1927, was removed, leaving only the floors and ceilings intact. The building was dedicated on December 11, 1978. It provided approximately 100,000 square feet of office space, including a board room, auditorium, seminar room, employees' lounge and commercial kitchen. CFC was able

to reduce its costs by leasing a portion of the building, reserved for future needs, to other businesses. The renovation of the Smith Building and the construction of an underground parking garage and a plaza cost approximately $5 million, giving CFC Square a total cost, excluding the residential space, of approximately $9.5 million.

CFC Square served CFC well until its sale at the top of the real estate market in 1989 for $21.5 million.

Charles B. Gill
Governor of CFC, 1979–1995

J. K. Smith walked into Clyde Denton's office one afternoon and sat down. "I've just hired my successor," he said.

The new employee was Charles B. Gill.

When Gill joined CFC as borrowers operations officer in April 1972, he was well-equipped for the challenge of helping member systems understand the still-new organization and developing new programs and services to help the members be more efficient and effective in their operations.

Gill began his career in rural electrification in 1961 when he joined REA as a trainee after his graduation from Duquesne University with a bachelor of arts degree in economics. He was quickly promoted to a succession of responsible positions, including field representative, chief of the Borrowers Management Branch, assistant director of the Office of Program Analysis and special assistant to the administrator. His experiences gave him insight into the challenges of managing a cooperative system and into the difficulties facing REA in terms of obtaining adequate loan funds.

While at REA, Gill received an appointment as a Presidential Fellow at the Graduate School of Economics at the University of Maryland. He completed all course requirements for a master of arts degree in economics.

As borrowers operations officer for CFC, Gill developed some of the organization's most successful programs. He organized and directed all field activities and was responsible for the administration of CFC loan security documents. He developed a series of cash management workshops that helped member systems increase revenues without raising rates and encouraged better financial planning.

In August 1978, Gill was promoted to deputy governor, a signal of the board's intent to name him as chief executive officer. He became governor in 1979. During his tenure, CFC's assets grew from $1.3 billion at the end of fiscal year 1979 to almost $7 billion at the end of calendar year 1994. The guarantee programs grew from $356 million at the end of fiscal year 1979 to $2.6 billion at the end of fiscal year 1994.

Gill is widely credited with developing new

opportunities for rural electrics to participate in innovative financing alternatives. "J. K. Smith had the entrepreneurial genius to pull the different elements together and to select the right people to manage this business at its inception," said Robert R. Edmiston, managing director of Lehman Brothers. "The genius of Chuck Gill was to take the organization to its next level, to turn it into a sophisticated broadly based service provider to the rural electric program and to develop and build an organization and array of services that met the needs of the membership. Yet Gill was able to do it in a fashion that did not create a bloated organization. He did not create a bureaucracy. He has kept the organization focused and member oriented."

Gill was responsible for the creation of the National Cooperative Services Corporation (NCSC), a financial service organization established by CFC in 1981 to offer new financing alternatives, including leveraged leases. He supported and encouraged the development of the National Rural Telecommunications Cooperative (NRTC) to assist member systems in meeting telecommunications needs in rural areas. He was also responsible for sharing the benefits of the CFC concept with rural telephone consumers through the Rural Telephone Finance Cooperative (RTFC), an affiliate of CFC created in 1987 to provide financing to rural telephone systems. He was appointed chief executive officer of RTFC at its inception.

Gill served on the board of directors of NCSC, NRECA International Foundation, RTFC, the Guaranty Funding Cooperative and NRTC.

Gill's quick mind served the CFC organization well during periods of high interest rates and in resolving difficult situations involving CFC members. "One of his strengths is his quick assessment of situations, his ability to comprehend a particularly challenging situation and see the important aspects of it without needing to go through a lengthy analysis," said CFC Senior Vice President and Loan Officer Richard B. Bulman. "He is able to understand quickly what issues would be important to a member, to the board of directors of CFC and to CFC as a company."

Gill's foresight has helped him leave the organization well-prepared for the future. "He has a really great ability to look down the road," said David J. Hedberg, CFC's director of rates and special projects. "He provided a vision to our membership. He made the tough decisions on how to position CFC to maximize its benefit to the members in the future."

Because of Gill's efforts, CFC and the rural electric program are much stronger in 1994 than they were in 1979.

first orders of business was to draw lots to divide the board into three groups to establish a pattern for director turnover. One group agreed to stand for reelection in one year and one in two years, and one agreed to serve a full three-year term.

The turnover of the first board was completed in 1977 when the first directors to serve two full three-year terms were replaced by eight newly elected directors plus the new president of NRECA.

The transition to new leadership included CFC's most senior official. In August 1977, J. K. Smith informed the board that he planned to retire at the end of 1979. He met with CFC's Operating Review Committee to discuss how CFC management continuity could best be maintained. The board agreed to establish a special position of deputy governor to be filled by a person qualified and expected to become governor. The deputy governor would serve in that capacity for a period of time. Then, assuming the board approved, the deputy governor would become governor and Smith would serve as governor emeritus until the end of the year.

Governor Gill

The Operating Review Committee selected the New York firm of Korn/Ferry International to assist in its search for candidates for the deputy governor position. On May 24, 1978, the board interviewed three candidates for the position. Charles B. Gill was appointed deputy governor. He succeeded J. K. Smith as governor on September 1, 1979.

Smith and Gill were alike in many ways. Both had a good grasp of the rural electric program and CFC's role in it. Both were strong, decisive leaders. Smith was also an entrepreneur and an innovative thinker. He was confrontational and demanding. "People were not ambivalent about Smith," Walsh said. "My experience was that they either liked him strongly or disliked him equally strongly. If you disagreed with him philosophically, he didn't do or say things that would make you like him any more."

Katzin observed that these qualities were needed at the time. "If an indecisive person, one who couldn't say no to anybody, had been in charge of CFC in its formative stages, it would have fallen apart," he said. "There would be competing organizations. There would be demands that were met that ended up being uneconomic or dangerous to the well-being of the organization."

By the time Gill was selected as governor, CFC was well established. Its needs were different. "Nobody can be a man for all seasons," Walsh said. "J. K. was the person for his season. His season was one of getting people to band together to make the initial equity investment in CFC; to hire the butchers, the bakers and the candlestick makers to make it work; to find us a place to live so we could own enough square footage that we didn't have to worry about outgrowing it; and to set up the arrangement between the CEO and the board of directors with very broad and meaningful delegations from the board to the CEO and make that work. It was building the organization and growing it to the point that it could stand on its own two feet, fighting for acceptance by the financial community and getting a good rating and then getting it upgraded. That was his job. He was a man for that season."

Gill was definitely the man for the next season. "Chuck was probably better equipped intellectually to handle the new problems that arose in the 1980s," Shesser said. "He brought a little more technical and economic background to his role. I think it has proven to be a very valuable asset in the light of the things that occurred."

"He's an economist, and he's a good economist," Katzin said. "He has a remarkable ability to conceptualize things."

Gill became governor at the beginning of what was probably the most difficult period in CFC history. His intellectual, conceptual and leadership abilities were about to be put to the test.

PART 4 CFC MATURES

During the first half of the 1980s, the loans, guarantees and other services CFC offered its members became much more sophisticated. It positioned itself to meet future program needs well into the 21st century by developing and implementing an equity stabilization plan. It enhanced its services to G&T systems and organized the National Cooperative Services Corporation to assist members in special financings such as leveraged leases. It expanded its ability to assist members in maintaining a strong financial performance by offering enhanced rate and regulatory analysis services. It instituted an associate member program for nonelectric utility services. CFC also surmounted critical challenges in the form of high interest rates and faced the emergence of a few troubled situations, including the bankruptcy of Wabash Valley Power Association.

Highlights

- *The National Cooperative Services Corporation (NCSC) was incorporated on February 10, 1981.*

- *The board established the Committee on Objectives and Planning in November 1981 to make an in-depth study of CFC's structure and operations.*

- *Between 1981 and 1986, NCSC provided more than $1 billion in debt to facilitate CFC member participation in leveraged lease transactions, including four major power plant leases, saving millions of dollars for cooperative consumers.*

- *In 1982 and 1983, NCSC assisted CFC members in 16 safe harbor leases of assets totaling $1.22 billion.*

- *CFC first offered a variable interest rate loan program in March 1983.*

- *CFC approved its first loan to an associate member in May 1983.*

- *The first pooled series of pollution control bonds guaranteed by CFC was issued in December 1983.*

- *More than 96 percent of CFC's members supported the equity stabilization plan by agreeing, before October 1, 1984, to extend the maturity of Subscription CTCs for an additional 50 years.*

- *Moody's upgraded the rating on CFC's collateral trust bonds to Aa2 from Aa3 in October 1984.*

- *The CFC board eliminated all late payment interest and premium charges on Subscription CTCs to encourage the remaining nonmember systems to join CFC.*

- *Member investments in CTCs exceeded $1 billion for the first time in 1985.*

When CFC was created in 1969, its functions and operations were relatively simple. It made long-term concurrent loans with REA and provided short-term financing. It financed these programs initially with its members equity contributions and later with collateral trust bonds and bank borrowings.

A decade later, CFC's world was much more complex, and its programs and services evolved to reflect that complexity. The conservation efforts engendered by national energy policies resulted in lower than projected growth rates, but the rural electric program was in the middle of a massive construction program undertaken when electricity needs were growing at a more rapid pace. The environmental impact of electric generation was a national concern, and the cost of pollution control equipment substantially increased the cost of new plants under construction. The word *troubled* was appearing with increasing frequency before the phrase *nuclear power plant,* but many new G&Ts were actively pursuing joint participation with investor-owned companies in nuclear projects as a way to gain ownership and a measure of control over their generating resources. The inflation rate hit 13.3 percent in 1979, and a double-digit rate was predicted for 1980. Nationally, interest rates reached the highest level since the Civil War.

On the positive side, the open animosity between cooperatives and investor-owned utilities had abated. "The ready access to financing the guarantee program provided to the co-ops, while the investor-owned companies faced deteriorating bond coverages and overcommitments in new plants, also opened the way for the G&Ts to buy into major projects under construction," Jerome Katzin said. "If a co-op could be encouraged to get its own generation and relieve the company's customers of costly new plant investment, so much the better. So much the better too, if the co-op was replacing an old, low, average cost source from the power company with a spanking new plant built at today's high costs with today's high-cost capital."

As utilities, electric cooperatives were particularly vulnerable to inflation. The resulting high interest rates substantially increased the cost of money. It brought higher operating costs for fuel, labor and other expenses. It increased capital costs for the new plants the cooperatives were building.

"The electric utility industry had become a much more difficult business in which to function," Ira Shesser said. "Up until the 1980s, running an electric utility business was complicated, but it dealt with issues that were well-defined, and people knew how to meet them." In facing these challenges and supporting its members' efforts to respond to them, CFC matured in its leadership role and in its ability to provide valuable services and financial programs.

A part of the maturing process that took place at CFC during the early 1980s was preparation for CFC's future. CFC had become much more than anyone had foreseen when the Long-Range Study Committee acted to create a cooperative self-help financing organization, but it retained the basic principles and functions established at that time. It was a supplemental financing organization, not a replacement for REA. It was a cooperative, and more importantly, it functioned as a cooperative in the way it treated its members and by refunding patronage capital. It recognized that it must be responsive to the needs and desires of the rural electric systems.

CFC Takes a Look at Itself

After 12 years of operation, it was time for CFC to take a fresh look at itself. "I'm a firm believer in the necessity for any organization to look at itself periodically and to develop a long-range plan regarding where it's going. That is far better than reacting to crisis conditions," Charles Gill said. "We wanted to bring the members in and encourage them to take a hard look at the organization to determine if any changes were necessary."

Gill believed that CFC had a responsibility to look beyond the present needs of the systems to make sure that CFC was prepared to meet the future challenges and requirements of the member systems. In the years of its operations, CFC had received many suggestions from its membership concerning its philosophy, organization and services.

In November 1981, the CFC board established the CFC Committee on Objectives and Planning to consider these and other issues. The board directed the committee to make an in-depth study of CFC's structure and operations to determine the extent to which CFC was meeting current member needs, to anticipate future member requirements and to recommend approaches to meeting those requirements. The board asked the committee to develop specific recommendations to guide CFC in its role of meeting the financing and financial service requirements of the rural electrification program in the future.

The committee followed the now well-established pattern set by the Long-Range Study Committee. James A. Vann, Jr., general manager of Dixie Electric Cooperative in Union Springs, Alabama, and CFC director from District 3, was appointed chairman of the committee. Twenty-five additional committee members were selected to represent a cross section of the rural electric program. There were two members from each of the 10 districts, and five at-large committee members.

The committee coordinated its work closely with NRECA's Committee on Financing for the Future, which was established at about the same time. The two committees held joint open-forum hearings in Minneapolis, Denver, Dallas and Atlanta. Approximately 600 people participated in the hearings.

The Committee on Objectives and Planning divided itself into three subcommittees to consider

- capital requirements and funding,
- objectives, organization and operations and
- services and information.

A New Look at Cooperative Attitudes

The Services and Information Subcommittee solicited member input on various issues through a confidential survey mailed to more than 1,000 rural electric systems. A total of 777 systems responded, including 671 member distribution systems, 35 nonmember distribution systems, 46 member G&Ts, one nonmember G&T, 21 statewide organizations and three service organizations. The questionnaire gave members an opportunity to express their views on CFC's organization, services and operations as well as on certain program issues. The results provided a quick view of rural electric attitudes at the time.

Most of the comments about CFC programs and operations were favorable. However, 61.5 percent of the respondents wanted to see CFC directors elected at the district meetings rather than the annual meetings. About half of the systems said rural electric cooperatives should be involved in nonelectric community assistance and development activities, but only 30 percent favored expanding CFC's membership to nonelectric rural utilities. Slightly more, 37 percent, said CFC should have loan programs

Members of the Committee on Objectives and Planning evaluated CFC's operations in the 1980s.

James A. Vann, Jr.,
Chairman
Dixie Electric Cooperative
Union Springs, Alabama

District 1
Charles C. Jones, Jr.
Virginia-Maryland-Delaware Association of Electric Cooperatives
Richmond, Virginia

Johnnie R. Austin*
Blue Ridge Electric Membership Corporation
Jefferson, North Carolina

District 2
Paul E. Weatherby
Cobb Electric Membership Corporation
Marietta, Georgia

Robert W. Williams, Jr.
Pee Dee Electric Cooperative, Inc.
Darlington, South Carolina

District 3
Robert C. Wade
Nolin Rural Electric Cooperative Corporation
Elizabethtown, Kentucky

Emmett H. Murray
East Mississippi Electric Power Association
Meridian, Mississippi

District 4
Edward P. Martin
Wabash Valley Power Association
Indianapolis, Indiana

Robert L. Roberts
Pioneer Rural Electric Cooperative, Inc.
Piqua, Ohio

District 5
Thomas H. Moore
Association of Illinois Electric Cooperatives
Springfield, Illinois

J. Bruce Bosworth
D. E. K. Rural Electric Cooperative
Estherville, Iowa

District 6
Virgil H. Herriott
Sioux Valley Empire Electric Association, Inc.
Colman, South Dakota

Everett Thies
Wright-Hennepin Cooperative Electric Association
Hamel, Minnesota

*Replaced Thomas Cockerham.

District 7

Vernon C. Williams
Yampa Valley Electric Association
Hayden, Colorado

Edward Baldwin
Wyrulec Company
Torrington, Wyoming

District 8

Leland F. Leatherman
Arkansas Electric Cooperative Corporation
Little Rock, Arkansas

John K. Davis
Sho-Me Power Corporation
Marshfield, Missouri

District 9

Byron Wagner
Big Bend Electric Cooperative, Inc.
Ritzville, Washington

C. R. Thiessen
Lower Yellowstone Rural Electric Association, Inc.
Lambert, Montana

District 10

Stanley K. Bazant
Plains Electric G&T Cooperative
Albuquerque, New Mexico

Alton E. Rollans
Concho Valley Electric Cooperative, Inc.
San Angelo, Texas

Members-at-Large

William B. Kittrell
Meriwether-Lewis Electric Cooperative, Inc.
Centerville, Tennessee

Frank Linder
Dairyland Power Cooperative
La Crosse, Wisconsin

Lewis E. Mitchell
Wheatland Electric Cooperative, Inc.
Scott City, Kansas

Clyde Rodolph, Jr.
Western Farmers Electric Cooperative
Anadarko, Oklahoma

Arnold E. Smith
Suwannee Valley Electric Cooperative
Live Oak, Florida

for nonelectric facilities.

Residential retail rates were higher than that of neighboring utilities, said 48 percent of the respondents. Among the regulated systems, 90 percent said regulation had a detrimental impact, with 33 percent seeing a severe detrimental impact. This revealed a significant change in attitude toward regulation. In comparison, only 12.4 percent of regulated respondents to the Long-Range Study Committee survey in 1968 reported unsatisfactory experiences with regulation.

Visitors to CFC Square enjoyed locating their cooperative on Roberto Morassi's giant sculpture hung in the lobby of the J. K. Smith Building.

Systems continued to rate availability as more important than the cost of money, with 56 percent of the distribution systems and 64 percent of the G&Ts (compared to 62 percent of the total respondents to the Long-Range Study Committee's questionnaire) ranking availability first. Although 26 percent of the Long-Range Study Committee respondents rated cost of money more important, 37 percent of the distribution systems and 31 percent of the G&Ts now rated cost as more important.

The rural electric program had not collectively considered the question of its basic philosophy since approving a statement of rural electric viewpoints and objectives as a part of the final report of the Long-Range Study Committee. The Committee on Objectives and Planning asked Virgil Herriott, who chaired the Objectives Subcommittee for the Long-Range Study Committee, to chair its Objectives, Organization and Operations Subcommittee to review the statement and consider whether there should be modifications in light of the changes in the rural electric program since that time.

Projecting Capital Requirements

The Capital Requirements and Funding Subcommittee projected future capital requirements for both distribution systems and G&Ts. It predicted that by the year 1990, the distribution systems would need between \$1.95 and \$2.16 billion of new debt each year, depending on changes in the REA program and the earnings levels of the systems. By the year 2000, distribution systems would need between \$4.64 and \$5.65 billion of new debt each year.

The projection for G&Ts, which was based on a survey of the G&T systems and planned capacity additions, predicted the annual need for funds for generation additions ranging from a low of \$452 million in the year 2000 to a high of \$5.3 billion in 1983. Nongenerating additions ranged from a low of \$402.3 million in 1990 to a high of \$1.4 billion in 2000.

How the rural electric community might best meet those needs was a major focus of the committee's deliberations. Experts from the financial and banking communities and government officials met with the committee to share their views. The committee also asked Lehman Brothers Kuhn Loeb, Inc. (the successor to Kuhn, Loeb & Company), to analyze the anticipated financial requirements of the distribution cooperatives and the G&T systems, to place the requirements in the context of the general capital markets and the competitive demands for investment funds and to evaluate the relative merits of private financing alternatives under varying market conditions and levels of government support. The committee then engaged Theodore Barry & Associates to review the Lehman Brothers study.

Lehman Brothers based its analysis on projected total capital requirements for distribution systems and G&Ts. The study noted that the capital requirements of the rural electric program would continue to be large, even with a nationwide reduction in the rate of growth. "The support of the REA has been essential to the success of rural electrification," it said. "The REA guarantee–FFB funding program has proven, in our judgment, to be a unique and irreplaceable capital resource for G&T systems, and essential to the recent very dramatic expansion of the G&Ts." The report went on to say that the private financing that had been obtained could not have been possible without strong investor confidence that REA would continue to play a dominant role in financing power supply systems.

"CFC is capable of enlarging its role in providing supplemental financing for distribution cooperatives to the extent necessary," Lehman Brothers concluded. "Even the fullest possible exploitation of available private financing alternatives would fall short of meeting the G&T requirements. It is particularly important to stress that even the perception of substantially diminishing REA support would limit the full utilization of available private-market alternatives. Access to the private market will be maximized if a strong on-going REA financial and administrative commitment is assured. Given this assurance, we believe that G&T systems have several alternatives for expanding their access to private capital."

The study ranked several alternatives according to cost-effectiveness. First was the continuation of the REA loan guarantee program with funding by the FFB. Second was the development of an agency status, similar to that of federal agencies such as the FCS, the Student Loan Marketing Association (Sallie Mae) and the Federal National Mortgage Association (Fannie Mae), which provided funding for various national programs. Those agencies received high bond ratings and were able to raise funds at the lowest cost. Third was the creation of a new Power Supply Financing Cooperative, which would operate in close coordination with CFC using CFC staff to carry out its activities. The study also evaluated raising G&T funds with a CFC guarantee and independent financing by individual G&T systems.

"CFC can play a vital role in G&T financing while continuing its traditional role with its distribution members," Lehman Brothers said.

The study also recommended that CFC further strengthen its financial position. "Regardless of the extent to which CFC expands its responsibility for G&T funding, it should strengthen its equity base by extending the duration of the Subscription CTCs and lowering the cost of the Loan CTCs," it said.

Recommendations of the Committee on Objectives and Planning

The final report and recommendations of the Committee on Objectives and Planning were presented to the CFC membership at the annual meeting in February 1983.

The committee found that the basic philosophy of the rural electric program had changed very little since the time of the Long-Range Study Committee. The few changes recommended in the statement of rural electric viewpoints and objectives reflected changes in the rural service areas, energy supply and utilization and economic conditions. For example, the objective on electric power usage and energy resources was expanded to recognize the need for wise development of all domestic fuel resources and the efficient utilization of electric services, not just the promotion of electric service.

The committee also recommended an addition to the viewpoints, one that perhaps did not need to be stated in 1969: "We believe: That all rural electric cooperatives must continue to work together in support of common goals and purposes to achieve their objectives effectively and efficiently." The committee recommended that each rural electric system consider and adopt the statement of rural electric goals and viewpoints.

In addition, it recommended that rural electric systems maintain a strong, unified approach to political, financial and economic efforts at the local, state and national levels.

In the months after the presentation of the committee's report, CFC area representatives contacted each of the cooperatives that was not a member to encourage them to join. The area representatives found that the nonmembers regarded the late payment interest and penalties on subscription CTCs, which could be quite large, as a significant barrier to membership. In comparing potential loans from CFC with those of other lenders, such as the Banks for Cooperatives, the systems often included the costs of joining CFC in the cost of the loan, putting the CFC loan at a substantial disadvantage.

Many of the systems that did not join CFC initially were not then required by REA to obtain supplemental loans. By 1984, REA required all borrowers to obtain supplemental loans. "In recognition of the changes taking place in the rural electrification program, CFC should give consideration to adopting a new approach for bringing nonmembers into membership," Loan Officer Gary A. Gordy told the board. "CFC's position as *the* supplemental financing source in the program would be significantly strengthened by having the greatest possible number of rural electric systems as members. With the uncertainty facing the entire rural electric program regarding federal provision of funding and support services, CFC must present itself as strong, stable and willing to act in behalf of the entire program."

Membership Made Easier

In November 1984, the CFC board took a significant step toward healing the rift that had existed between CFC and many of the nonmember systems by eliminating all late payment interest and premium charges on Subscription CTCs. Furthermore, it approved the refund of premiums and late interest charges collected from systems that joined after the initial subscription period.

Sheldon Petersen was the area representative serving North Dakota and South Dakota, where many of the systems that did not join CFC were located, when the board took this action. "I had one system in North Dakota that had elected to become a CFC member just a few days before this happened," he said. "One of the neatest things I ever did was to call the manager and tell him he didn't have to pay the penalties. He was ecstatic, because it reinforced his decision." And he saved over $80,000.

"Once CFC was organized and had the broad-based support that we had, it was time to take a more global picture and look at the program benefit as a whole," Petersen said. "There is a benefit to a new member who comes in under favorable arrangements, but it is also a benefit to the existing members for CFC to be able to represent itself as the representative of the rural electric cooperatives in the area of finance."

In other areas, the Committee on Objectives and Planning recommended three changes in CFC's bylaws, which were presented at the 1983 annual meeting. The first was a recommendation to elect CFC directors at the district level rather than in a national election. "That had been a controversial subject for some time," Gill said. District 7 had formally proposed that change as early as 1971. Most of those voting favored the amendment, but it failed to receive the necessary two-thirds majority to pass.

The second proposed bylaw amendment would allow CFC to establish an associate membership for not-for-profit corporations owned, operated or controlled by rural electric systems eligible for Class A (distribution system) membership in CFC. It was envisioned that the associate members would provide nonelectric services to rural areas, such as water and wastewater service, telecommunications, and health and safety services. The amendment passed, and in March 1983, the board approved the procedures for implementing an associate member program.

The committee also recommended a technical amendment that would allow CFC to reduce the level of contributions to its education fund. The statute under which CFC was organized required a portion of its net margins to be used for cooperative education. In 1980, CFC established a program to provide grants from that fund to statewide organizations to support specific education programs. The bylaw amendment, which passed, reduced the annual contribution to the education fund from 1 percent to 0.25 percent of net margins. At this level, CFC could still provide a meaningful education program while reducing the impact of the contribution on CFC's TIER and patronage capital refunds. From 1980 through 1994, CFC distributed more than $1.5 million in grants from the education fund.

Strengthening CFC's Equity Base

The most significant of the committee's recommendations were those that related to strengthening CFC's equity base. The suggested changes were aimed at stabilizing the equity for the long term, equalizing members' equity contributions and ensuring CFC's borrowing capability through a strong debt-to-equity ratio.

When CFC was started, the distribution and power supply members agreed to provide equity capital on an annual basis from 1970 to 1984 through purchase of Subscription CTCs. Systems were required to purchase additional Loan CTCs as a condition of loans. The Subscription CTCs were scheduled to mature in 2020, 2025 or 2030 depending on the year in which they were issued. Because most of the long-term loans made by CFC had a maturity of 35 years, the loans made in 1985 or later would last longer than part of the equity supporting them. In addition, the Subscription CTCs were the only source of equity for CFC's short- and intermediate-term loan programs, which did not require purchase of CTCs.

The committee recommended that CFC develop a plan to extend the term of the Subscription CTCs or to make them permanent in nature to stabilize CFC's equity base. The committee also recommended that CFC consider various changes in Loan CTCs.

In March 1983, the board approved changes in the Loan CTC policy that were effective on June 1, 1983. Under the new policy, a member could choose to include the full amount of the Loan CTCs in the loan and make that investment with the first advance on the loan or to make the investment from general funds in installments over 12 quarters beginning with the first quarterly debt service billing. The CTCs would be amortized over the term of the loan and would no longer earn interest. The amortization would maintain the CTC investment at approximately 5 percent of the loan balance. The overall effect of these changes was to reduce the cost of loans to member systems.

The board asked Lehman Brothers to provide additional analysis on alternatives for the Subscription CTCs and the impact on CFC's cost of funds, leverage, margins and performance ratios. Based on that analysis and other information, in August 1983, the board asked the CFC members to voluntarily extend the maturity of their Subscription CTCs for an additional 50 years. In return, CFC would increase the interest payments on the subscription CTCs from 3 to 4 percent from October 1, 1985, through September 30, 1992 (one full 7-year loan repricing cycle), and to 5 percent starting October 1, 1992. Members who did not choose to extend their CTC maturity would continue to receive the original 3 percent interest rate. The members were asked to notify CFC of their decision by October 1, 1984.

Over 96 percent of CFC's members supported the equity stabilization plan by extending the maturity of Subscription CTCs.

"Essentially, the Committee on Objectives and Planning went back to the members and said, 'Tell us again whether CFC is the right organization for the rural electric program,'" Gill said. "I was uplifted by the tremendous loyalty shown by our members. This comprised a kind of rededication to the principles of CFC—principles that had been established throughout the history of CFC. This rededication was demonstrated when all but 34 of our members agreed to extend their CTCs for an additional 50 years. That's the biggest thrill I received from the entire project." By 1994, fewer than half of those 34 systems still had not extended.

High Interest Rates, Tight Money Supply Present Challenges

The economic conditions of the late 1970s and early 1980s had roots in the federal government policy of the mid-1960s. "The Johnson administration had a guns and butter policy instead of a guns or butter policy," said Robert Edmiston, managing director of Lehman Brothers. "They did not implement tax increases or spending cuts and continued domestic spending at the same level while overlaying the costs of the Vietnam War."

Then, in the late 1970s, there was an increase in the money supply. The federal deficit was out of control. Interest rates were below the level of inflation. The Federal Reserve Board, concerned that the inflation situation could get even worse, began to make money much less available to the economy. "It pushed short-term interest rates up dramatically," Edmiston said. "That precipitated the recession of 1980 and 1981. Interest rates went up very high, because people were more sensitized to inflation. They were uncertain as to whether the policy of the Federal Reserve was going to be successful." Treasury rates on 30-year

Robert R. Edmiston, managing director of Lehman Brothers, addressing the board at CFC's Herndon facility, reported frequently on economic conditions to help the board evaluate the market for new bond issues.

bonds reached the 14.7 percent level in October 1981, and there was an inverted yield curve, that is, short-term interest rates were higher than long-term interest rates. In fact, three-month Treasury bills yielded 15.5 percent, and the bank prime lending rate was 20.5 percent. "All of this was necessary to begin to bring to bear some discipline on the growth of the money supply and thus on the rate of economic growth and the level of inflation. The Federal Reserve managed to set in place the sequence of policy actions that brought inflation under control gradually during the 1980s. Even today [1994], though, the bond markets are very, very sensitive to inflation," Edmiston said.

CFC faced some difficult decisions in issuing bonds during that period. The board had established a policy to issue long-term bonds when about $50 million of short-term debt was being used to fund long-term loans. "The issue we had to deal with when interest rates began to rise was that CFC had delayed financing by making loans out of its equity base in anticipation of rates coming down. In fact, rates went much higher," Edmiston said.

CFC had always included an optional call provision, which would allow it to repurchase its bonds before the scheduled maturity date. "If rates were going to be that high, we decided we had better consider shortening the maturity on the bonds, too," Paul Fox said.

This generated a great deal of discussion among the board and finance committee members. "CFC's debt capital to support long-term loans should have a mix of long-term and intermediate-term maturities," Gill told the board in May 1979. Although there was not a specific policy on the issue, staff recommended a ratio of about 60 percent long-term debt to 40 percent intermediate-term debt.

Series F and G, issued in 1979 at 9.75 percent, were both 30-year issues with early redemption options after 10 years, although there was considerable discussion about issuing intermediate-term securities instead because of the "high" level of interest rates. "Those turned out to be cornerstone issues, given the high interest rate environment of the 1980s. When CFC bond interest rates were in the 15 percent range 2 years later, those transactions obviously looked very attractive," Edmiston said.

Series F and G were followed in 1980 by Series H at 12.625 percent and Series I at 13.5 percent and in 1981 by Series J at 14.75 percent, Series K at 14.875 percent, Series L at 15.75 percent and Series M at 14.625 percent. Series H, K and L had 10-year maturities, whereas Series M had a 7-year maturity. The others were long-term issues. "All of those transactions were very difficult to do," Edmiston said. "It was very difficult, even at those high interest rates, to sell the obligations, because of the enormous uncertainty created by a rapidly declining bond market. Interest rates were at unprecedented levels, and bond investors had substantial unrealized losses on major portions of their portfolios. Indeed, by the third quarter of 1981, the 9.75 percent Series F and G bonds were trading to yield about 15.75 percent, making the bonds worth about 62 cents on the dollar.

Can you imagine buying a high-quality, double-A credit, and two years later, without any change in the credit rating, finding you have lost 38 percent of your original investment?"

The board's Bond Pricing Committee, consisting of the Operating Review and Audit Committee and two or three at-large members of the board, was responsible for making the decisions on bond issues. The pricing meetings were usually held at Lehman Brothers' offices in New York. The governor, finance officer and corporate counsel joined the committee members for an extended review of market conditions. Lehman Brothers provided a booklet with all the relevant data on the market for the issue, including an economic forecast, the trading range for outstanding CFC issues and comparable issues, and the volume of new issues in the market. "The object was to place the issue within the context of the overall market," Katzin said.

Then the discussion turned to terms. "Often the issuer had several trade-offs to consider, balancing the size of the issue, term, call provisions and other characteristics that affected marketability and all of which bore on the overall cost," Katzin said. Then the committee decided whether to go ahead on the terms that could be worked out or to postpone the offering. In the early 80s, the decision frequently was to postpone. "We then maintained continuous contact, talking every day or so about the moves in the market and the outlook. At some stage, the decision was made to try again, and the procedures were repeated."

Bond Pricing Committee Makes Difficult Decisions

The decision to market Series M illustrates the level of consideration and concern that went into each of these decisions. The series was registered with the SEC on October 23, 1981, so that CFC would be in a position to price the bonds quickly when market conditions became more favorable. Governor Gill asked the Bond Pricing Committee to convene in Washington to review the situation on Thursday, November 18, during a regularly scheduled board meeting. Because the board was in session, the committee asked that the proposed sale be reviewed with the full board for its input.

Gill reported that, if the bonds were sold on that date, the price for a 10-year issue would be between 14.625 and 14.75 percent, for a 7-year issue between 14.5 and 14.625 percent, and for a callable 30-year issue approximately 15.75 percent.

He pointed out to the board that CFC had approximately $160 million in long-term collateral that needed to be backed by long-term funds within a few weeks. In addition to the lower rate, the 7-year bonds could be sold with a 5-year no-call provision, which would allow CFC to refund the bonds during the sixth and seventh years at par, assuming more favorable interest rates at that time.

Edmiston reported on current market conditions, but the board asked him to get further information from Lehman Brothers' fixed income marketing group. At Edmiston's request, that group held a special session to review once again the market and competing issues and to provide their up-to-the-minute input to the Bond Pricing Committee.

Having obtained information from all the sources available to them, the Bond Pricing Committee decided to go with the seven-year issue. It was priced at 14.625 percent after a 14.5 percent offer was rejected by the market.

CFC then positioned itself to issue Series N when a favorable window presented itself. Over $100 million in long-term loans were being "warehoused" with short-term debt while CFC waited for the market to improve. Fox reported to the board in May 1982 that alternatives at that point included a 30-year issue priced at about 15.875 percent, a 10-year issue priced at about 15 percent and an issue of zero-coupon serial bonds (bonds that pay interest only at the maturity date). The board adopted an amendment to its resolution authorizing Series N to allow the new zero-coupon bonds, which were becoming attractive to investors, as an option. Fortunately, CFC was able to negotiate a $500 million bank revolving credit agreement that alleviated the pressure to issue bonds at that time. The Bond Pricing Committee reviewed market conditions and CFC's financial situation frequently, but CFC was able to delay Series N until 1985. By then, the rates had ameliorated substantially, and the 10-year issue was priced at 10.5 percent.

Commercial Paper Keeps Rates Down

Commercial paper sales were an important source of capital during this period. At the peak in fiscal year 1981, CFC commercial paper provided $1.3 billion in capital to meet rural electric needs, including $969 million in dealer sales, $301 million in direct sales to members and $7 million in sales to nonmember systems. Of course, the cost was at its peak, also. The average interest rate on commercial paper during fiscal year 1981 was 16.78 percent.

Despite the high interest rates, CFC's strong equity base enabled it to offer rates to its member systems that were below the market at the time. For example, during the time when CFC's cost of long or intermediate-term funds was greater than 15 percent, the maximum interest rate on CFC's standard long-term loan, which would be repriced after seven years, was 14 percent. "We were careful to educate our members as to what we were doing and why we were doing it," Ira Shesser said. "We were able to assure them that we could continue to provide them with funds during this difficult period. However, they would have to pay the freight on them." CFC had to balance maintaining rates as low as possible for its members with its own financial needs. "I felt that CFC had to maintain certain financial standards that would make us acceptable to the financial community," Fox said. "Just to keep our head above water in the financial community, we had to shoot for a 1.10 TIER. That meant only one thing. We had to raise rates." He did an analysis and presented it to Gill, who said no. Member needs were the top priority, and the financial community would have to be satisfied another way. CFC's TIER fell to 1.06 in 1981. Despite that, because of its other strengths, CFC maintained its double-A rating throughout the period.

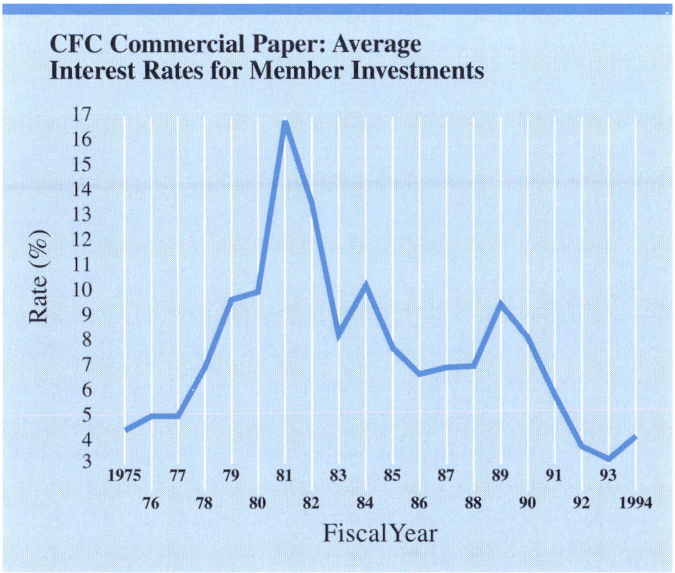

In the early 1980s, the commercial paper program helped CFC ride out the long-term bond market until rates returned to more acceptable levels and provided an attractive investment for member systems.

CFC made numerous policy changes during this period to make it easier for members to manage their borrowings. Initially, CFC's long-term loans were made at rates fixed for the entire 35-year period of the loan. CFC had amended that policy in 1974 to fix the interest rate on long-term loans approved after

September 16, 1974, for the initial 7-year period only. In the seventh year, a new interest rate was fixed for the next 7 years at the then-current rate for new long-term loans.

"That was a good move on CFC's part," Shesser said. "It gave us flexibility when interest rates went sky high. We were not hurt because we were repricing our interest rates. Out of that came other loan terms that are now in place."

In response to member concerns about the volatile interest rate climate, CFC added an option for systems entering the seventh year. In March 1982, the board adopted a policy effective in the repricing year beginning January 1, 1983, that allowed a system, at its discretion, to elect to take the interest rate into effect on January 1. That rate would be in effect for the entire seventh year. If CFC's interest rate increased during the year, the borrower would be assured of the January 1 rate for the next seven years. If rates declined during the year, the system would be given, as of the following January 1, the lowest rate offered during the year for the next seven years. The system could also choose to repay its CFC loan in full at the time of repricing.

CFC's initial long-term loan program established the interest rate at the time the loan was approved, but often, the funds were not drawn down by the borrower for up to two years. When interest rates began to decline, the CFC rate dropped from 14 to 11 percent between July 1982 and May 1983, less than a year. Members who had loans approved at the 14 percent rate on which they had not drawn funds understandably objected to the higher interest rate. CFC addressed the problem by offering new interest rate options and changing the loan pricing policy.

Although CFC philosophically preferred fixed long-term interest rates for long-term loans, it recognized that a variable-rate program could offer benefits in some situations. "The market forced us to change our thinking, and as a result, since that time, CFC has been very sensitive to market conditions. We've structured our borrowing and lending programs to reflect that," Shesser said.

In March 1983, the board approved a variable interest rate program that allowed any system having an approved loan against which funds had not been drawn down to convert the loan to a variable interest rate on payment of a conversion fee to cover CFC's cost of funds. The interest rate on the variable-rate loans was established monthly. The systems had the option of later returning to a fixed interest rate based on the cost of funds at the time. CFC also offered the variable interest rate option for new long-term loans and as an additional repricing option. In 1984, the variable-rate program was expanded to allow conversions of fixed-rate loans that had been advanced.

CFC further enhanced the flexibility of its loan programs in 1984 by pricing loans at the time of the first advance, rather than at the time of approval.

"The new options permitted CFC members to select the programs best suited to their financing objectives and philosophies," said Gary Gordy, who was CFC's loan officer during that period.

Economic Conditions Begin to Improve

The years 1983 and 1984 were pivotal for CFC financing. Interest rates began to fall. Fox left CFC to pursue other interests and in 1984 was succeeded as finance officer by Truman L. Brandt. "My number one objective was to try to get Moody's to give us an upgrade, because at that time their rating was below Standard & Poor's," Brandt said.

For the first time, CFC was funding very large G&T transactions involving hundreds of millions of dollars, so there were new credit questions to address. Credit concentration became an issue as well as the complexity of some of the transactions. "We needed to show the rating agencies what our objectives were," Brandt said. "Credit markets don't like surprises. They like stability, and they like to see you accomplishing your objectives and maintaining them. If you run into any surprises, you better have a good story, or you are going to pay the price in the marketplace."

A fortuitous event prompted CFC to reestablish a reserve for loan losses. CFC first established a reserve for doubtful loans targeted at 0.25 percent of total assets in 1971. The target was reduced to 0.125 percent in 1973. The reserve never reached a significant level. Because the risk of a loss was perceived as being very low, the board never increased the account on the basis that the equity in unallocated margins was sufficient to cover any losses and unnecessary contributions to the reserve would reduce margins and thus members' patronage capital. The entire reserve of $302,000 was abolished in September 1981.

In early 1984, the Northwestern Mutual Life Insurance Company, which had accumulated over $31 million of CFC's Series

A bonds, approached CFC about selling them back. On February 14, 1984, CFC purchased the bonds at 70 cents per dollar of face value. This resulted in an extraordinary gain of over $9.4 million. The indenture for Series A required CFC to make a series of annual mandatory sinking fund payments sufficient to redeem the principal amount of the bonds. Although, at 7.4, percent, the interest rate on the bonds was lower than CFC's current cost of long-term funds, the purchase met CFC's sinking fund requirements for Series A for the years 1989–2004. In addition, should CFC wish to increase its debt-to-equity leverage ratio, it could eliminate the 10 to 1 limit in effect under the Series A indenture (the limit was raised to 20 to 1 for subsequent issues) by retiring the relatively small balance of bonds outstanding after the purchase.

Despite difficult conditions in the capital markets of the early 1980s, CFC continued to achieve its financial objectives.

- Providing financing, financial products, advice and services at the lowest possible cost for electric and other utilities
- Strengthening ties to the financial community to improve acceptance in financial markets, liquidity, and availability of funds at favorable rates
- Improving bond ratings
- Maintaining times interest earned ratio (TIER) at about 1.10
- Maintaining a reasonable leverage ratio that supports business growth and is consistent with prudent management objectives
- Maintaining a strong, viable financial position to ensure confidence of the investment community
- Developing new financial service products to meet member needs
- Providing for a seven-year (or less) patronage capital retirement cycle
- Assisting members in maintaining adequate TIERs and adequate earning ratios and responding to other member needs

There was a question of what to do with the gain. "We thought it would be unfair to give the windfall as patronage capital to only those systems that had loans during that year," Gill said.

Brandt recommended that the gain be put into a loan loss reserve. "I was so new I hadn't had time to evaluate the credits," he said. "What I did know was that you can't have that many loans outstanding and claim that you don't have a potential exposure."

If there were no losses, the funds could be dissipated from the reserve back to the members over a period of time.

In March 1984, the board adopted a policy to establish an allowance for loan losses to be accumulated over time to 0.5–1.0 percent of loans outstanding. The board allocated $10 million to the fund for fiscal year 1984. It later added the $16 million gain on the sale of CFC Square and started to make significant annual contributions. The wisdom of this action became apparent as troubled situations emerged over time.

Protecting Against Risk

Although interest rates had begun to fall, there was a concern that they would go up again. At the end of fiscal year 1984, CFC had long-term fixed-rate loans outstanding totaling $1.9 billion. The total amount of collateral trust bonds outstanding was $900 million. Commercial paper outstanding totaled $1 billion, including $508 million supported by revolving credit agreements that was classified as long-term debt. The heavy reliance on commercial paper made CFC especially vulnerable to increases in interest rates.

CFC protected itself against that risk by entering into interest rate swap agreements. Basically, the swap allowed CFC to fix the rate on a portion of its commercial paper funds by exchanging a short-term interest rate with another company for a long-term rate. No additional funds were actually raised, and no principal changed hands. The first swap was executed on June 19, 1984, with Bankers Trust Company. CFC obtained an all-in cost of 13.9 percent on the $50 million swap, which had a seven-year term.

"The value of the swap was that we got an effective rate through the swap lower than we could by selling securities directly in the market place," Brandt said. It also provided a hedge against rising interest rates.

Effective October 9, 1984, Moody's raised the ratings on CFC's collateral trust bonds to Aa2 from Aa3. The upgrade was based on the improved financial condition of CFC's member systems as well as CFC's own performance. Moody's also cited the near unanimous support for CFC's equity stabilization plan and the establishment of the loan loss reserve. "Although CFC held an Aa3 rating for its collateral trust bonds, the perception with the upgrade to Aa2 is that CFC is now a solid Aa bond, and that gives the investor greater comfort," said Moody's Assistant Vice President Loretta Neuhouse.

While Lehman Brothers continued to serve as CFC's lead underwriter, the board added five investment banks to the approved list in November 1984. Lehman's leadership and performance in marketing CFC's credit continued to be outstanding, but with changes in the complexity and nature of CFC's financings, it wanted access to more firms. The additional investment banks included Goldman Sachs & Company, Merrill Lynch Capital Markets, Paine Webber, Inc., Salomon Brothers, Inc., and E. F. Hutton & Company. (By the 1990s, CFC had concentrated its work with Lehman Brothers, Goldman Sachs and Merrill Lynch.)

CFC also started meeting directly with investors as it had done with bankers. "We held meetings around the country with investment analysts and took our story directly to them," Brandt said. "That helped because it increased the demand for our securities. More people became comfortable with CFC. It increased demand for our securities and lowered cost."

Meeting the Needs of the G&Ts

By the end of fiscal year 1980, more than one-third of CFC's total loans outstanding of $1.8 billion and virtually all of the $592.8 million in pollution control bond guarantees were for G&T systems. CFC consolidated its efforts to serve the G&Ts by creating the power supply department on June 1, 1980. Richard Bulman, who had experience with both REA and the Southwestern Power Administration, was selected as power supply officer. "The idea was to provide what I would call today value-added services to the G&Ts and to maintain CFC's expertise at a high level so we could demonstrate to the rating agencies and our board that we knew what we were doing when we made these large commitments to the G&Ts," he said.

The power supply department focused on developing services that reflected the uniqueness of individual G&T systems. A new publication, the *CFC Power Review,* discussed power supply issues such as forecasting, fuel supply and environmental concerns in addition to new financing options.

Responsibility for CFC's annual Power Supply Symposium shifted to the power supply department. Discussion in the early 1980s focused on issues such as changes in the tax laws, leveraged leasing, tax issues related to nonmember income, national energy policy, marketing and environmental concerns.

During the period of volatile, high interest rates, CFC developed a program to estimate FFB interest rates on a daily basis—they had previously been available only after the fact—and to provide a risk-reward analysis of the break-even rate for long-versus-short-term drawdowns. This helped systems manage the timing and term of their drawdowns of FFB funds guaranteed by REA.

The KRTA was well-recognized as a valuable credit analysis and management tool for the distribution systems, but CFC had not offered a similar service for the G&Ts, primarily because of the uniqueness of each G&T system. CFC developed the G&T Trend to fill that void. By comparing individual G&T performance in 36 key areas with that of a designated peer group of G&Ts with similar power supply sources, the G&T Trend established a reasonable standard of comparison. And any G&T could request a comparison with a peer group of its own choosing. "To this day it has been a good reference vehicle internally, and it has been a good reference for member systems, too," said Martin R. Crowson, who joined CFC's special finance group in June 1982. As a related service, CFC compiled an annual comparison of wholesale power costs.

As the attention of investors and the rating agencies focused more and more on nuclear issues, not just for the rural electric program but for the entire utility industry, the power supply group was often called on to field questions about CFC's nuclear involvement. Cooperative systems ultimately acquired over 3,000 megawatts of operating nuclear capacity in 15 plants. The projects achieved varying degrees of success. CFC evaluated each member's needs on an individual basis and did not change its services to any member solely because of nuclear participation.

The power supply department merged with the loan department in 1984 when Bulman was appointed loan officer.

Financing Is the Thing

Although the services offered by the power supply department and later the power supply group in the loan department were well-received, the issue that was of most interest to the G&T systems was, of course, the actual loan and guarantee programs.

The line-of-credit and intermediate-term loan programs were major elements of CFC's G&T services throughout the first half of the 1980s. They were used primarily for front-end financing for projects under construction that already had long-term REA financing available. Typical of the time was the May 21, 1980,

Loan Committee meeting. The committee ratified the action it had taken by mail, approving 10 G&T lines of credit totaling $189.4 million and three G&T intermediate-term loans totaling $169.4 million. The committee approved six more G&T lines of credit totaling $107.4 million and recommended for approval by the full board eight G&T lines of credit totaling $331.7 million and four G&T intermediate-term loans totaling $325 million.

The intermediate-term loans requiring board approval included two to Arkansas Electric Cooperative Corporation (AECC) to insure continuation of construction on the White Bluff and Independence plants. AECC had a 35 percent ownership interest in the White Bluff station, two 700-megawatt coal-fired units being constructed by lead participant Arkansas Power & Light Company (AP&L), and a 35 percent ownership interest in the Independence station, two 700-megawatt coal-fired units also being constructed by AP&L. AP&L had notified AECC that it was unable to meet its construction financing obligations. The ownership agreement allowed AECC to make payments in behalf of AP&L to avoid construction delays and the attendant cost increases. If AP&L were unable to repay AECC, the ownership interests would be adjusted. There were several similar situations where the availability of CFC's resources facilitated construction on projects that might otherwise have been substantially delayed because of the lack of funds.

The favorable tax laws in effect during the early 1980s allowed CFC to offer many new programs benefiting the G&T systems, ranging from safe harbor leases to tax-exempt financings. There was an expansion of CFC's guarantee options for tax-exempt and other types of guarantees.

By 1982, CFC was one of the nation's largest non-insurance-based financial guarantors of tax-exempt securities. It was the first organization to guarantee tax-exempt commercial paper. "As a result, the members did not have to lock in fixed rates for the full term during a period of high interest rates. It allowed them to ride down the variable rate," said Steven L. Lilly. Lilly joined CFC in 1983 and became principal finance officer in 1993 after Brandt's retirement and chief financial officer in 1994. "Subsequently, they were able to lock in at a significantly lower fixed rate."

Further Savings with Tax-Exempt Pool

In September 1982, the CFC board approved another innovation in tax-exempt financing, the creation of a guarantee of a "pool" of tax-exempt securities. CFC's tax-exempt guarantee program was restructured to place the mortgage collateral for issues in a pool, thus providing diversity to investors, much as CFC's collateral trust bonds provided diversity. CFC hoped this would establish stronger security and lead to an upgrade of the bond ratings for this type of CFC-guaranteed issues.

The pooling concept was very complicated and took many months to develop. Beyond that, there was the task of getting the rating agencies to assign a higher rating to the initial transaction. "Their concern was that there wasn't yet enough diversity in the pool to support a higher credit rating and that there would be an undue concentration of risk," Robert Edmiston said. "They said they would not upgrade the credit until there were a large number of participants in the pool."

Edmiston researched modern portfolio theory, which analyzes risk diversification and the risk of loss, and found that the probability of loss decreased markedly when a pool had five or more members. Although one agency continued to wait for a larger pool, the other rating agency recognized the value of the pooled structure and provided a rating upgrade. "We were able to get Moody's to accept a higher credit-quality rating with 5 cooperatives in the pool instead of waiting for 10 to 15 to join because we were able to demonstrate to them that statistical portfolio analysis showed that the risk of loss fell off sharply with as few as 5 participants in a particular portfolio," he said. The bonds were initially rated Aa3 and upgraded to Aa2 in 1984 when CFC's collateral trust bonds were upgraded to Aa2.

The first pooled series guaranteed by CFC was issued on December 16, 1983. It included six issues totaling $141.3 million for five G&T systems. Like the other tax-exempt guarantees, this program allowed CFC's members to raise money at a more attractive rate than they would be able to do on their own. "Individually, none of them had the credit quality CFC had," Edmiston said.

Unfortunately, the Tax Reform Act of 1986 eliminated private activity bonds for pollution control equipment financing, limiting the scope of the program to resource recovery private activity bonds. However, CFC members continued to benefit from over $1.5 billion in lower cost financing obtained with the help of CFC's pollution control bond guarantees and standby liquidity.

Leases, Tax-Benefit Transfers Offer Economic Advantages

CFC's success in developing tax-exempt financing programs stimulated the G&Ts' interest in other forms of tax-advantaged financing. In 1979, CFC held a conference to discuss leveraged leasing and how cooperatives might participate in lease transactions. Although there were questions about the attitude of the IRS toward cooperative participation in tax-advantaged transactions and the rural electric systems lacked experience in leasing, it appeared to offer benefits.

"We need packagers to seek out equity participants and negotiate with them," one participant said. "In addition to that, we need in-house attorneys or our own legal firm working directly for the cooperative to provide legal guidance for these transactions. Also, either in-house financial personnel or outside consultants are needed. We need someone to evaluate what the packager is saying, someone to analyze very quickly the advantages of payments in advance or payment in arrears and many other similar matters."

The G&T Managers Association appointed a committee chaired by Robert N. Cleveland, manager of Buckeye Power Cooperative in Columbus, Ohio, to work with CFC, NRECA and REA to explore leasing options. The result of that effort was the incorporation of the National Cooperative Services Corporation (NCSC) on February 10, 1981. NCSC's initial purpose was to assist cooperatives in evaluating leveraged lease proposals and to enable them to raise debt capital for lease transactions at interest rates lower than those otherwise available.

It was necessary to create NCSC because, under the applicable tax laws, related parties could not provide or guarantee debt in a leverage lease transaction. CFC could not be the direct debt participant because of its policy to lend only to members. For those reasons, NCSC's members were distribution systems that were members of G&T cooperatives. NCSC joined CFC as a Class C member. CFC is also a member of NCSC. CFC contracted to provide management services to the new organization and agreed to guarantee NCSC debt. The fees charged for transactions plus interest income on the subordinated-term certificates purchased in connection with the CFC guarantees would provide revenue to meet NCSC's operating expenses.

CFC and NCSC expanded their relationships with some of the leading investment banks to provide expertise and additional support services as needed. The group included Lehman Brothers,

Paine Webber and Goldman Sachs.

In a basic leveraged sale lease-back transaction, a company anticipating a large tax obligation for several years purchased an asset from an organization that did not expect a large tax obligation during that same time. The purchaser, or lessor, who had no need for the actual asset, then leased it back to the seller. The lessor obtained the tax benefits associated with ownership of the assets, including accelerated depreciation, interest deductions, investment tax credits and energy tax credits. Those benefits were shared with the seller, or lessee, through the lease payments. Depending on the circumstances and equipment involved, leasing could provide a lower cost alternative to owning the equipment through debt capital as well as a hedge against obsolescence or inflation, and it sometimes provided access to capital that might not otherwise have been available.

Senior Financial Consultant Steven L. Lilly, who became senior vice president and chief financial officer in January 1994, discusses the potential benefits of leveraged leasing at a seminar for G&T representatives.

On the other hand, lessors often required lessees to indemnify them against changes in circumstances that would reduce their benefits. Furthermore, the lessee lost some control over the asset. The legal title to the asset was transferred to the lessor. At the end of the lease term, if the cooperative wished to regain ownership of the asset, it was required to repurchase it at the future fair market value. Alternatively, it could continue to lease the asset at a reduced rental rate.

Arranging a leveraged lease transaction was a costly, time-consuming project. NCSC generally recommended that it was not cost-effective for assets of less than $5 million.

In the proper circumstances, however, leasing could be a very attractive form of financing. By 1984, NCSC estimated that a total of $160 billion had been invested in leased equipment in the United States.

National Cooperative Services Corporation helped CFC members lease more than $1 billion in assets by arranging or participating in leveraged leases.

Date	System
December 1981	Associated Electric Cooperative, Missouri
June 1983	Western Fuels Association,Wyoming, for the benefit of Sunflower Electric Cooperative, Kansas
August 1983	Seminole Electric Cooperative, Florida
October 1983	Arkansas Electric Cooperative Corporation, Arkansas
December 1983	Western Fuels Association, Wyoming (Subleased to Western Fuels-Utah, Wyoming) for the benefit of Deseret G&T Cooperative, Utah
August 1984, November 1984	Cherokee Falls Association, South Carolina, a service contract for the benefit of Broad River Electric Cooperative, South Carolina
December 1984	Arkansas Electric Cooperative Corporation, Arkansas
December 1984	Seminole Electric Cooperative, Florida
December 1984*	Colorado-Ute Electric Association, Colorado
December 1985	Deseret G&T Cooperative, Utah
July 1986	Western Fuels-Utah,Wyoming for the benefit of Deseret G&T Cooperative, Utah
December 1988	Arkansas Electric Cooperative Corporation, Arkansas

*NCSC arranged for CFC's guarantee of debt service.

Debt Provided (in millions)	**Leased Item**
$18	Dragline
2.3	115 rail cars
41.8	Tugboats, towboats, barges, rail cars
6	Coal mine equipment
5.2	Locomotives and rail cars
9.1	Hydro project (dam)
148.6	Plant lease
225.9	Plant lease
	Plant lease
377.9	Plant lease
16.1	Deserado Mine equipment
71.6	Hydro project (lock and dam leases)

NCSC Provides Lease Debt

The leveraged lease transactions required both debt and equity investment. NCSC's ability to borrow funds from CFC and/or to apply for a CFC guarantee of its publicly issued or privately placed debt gave NCSC great flexibility in structuring lease transactions. "NCSC's primary role was to permit our members to execute sale–lease-back arrangements on power plants, rail cars and other transactions by acting as a funding conduit for CFC loans and guarantees," Lilly said. "NCSC was the transaction manager. We identified major investment banking firms that placed transaction equity and also underwrote the lease debt." Although the tax laws required a minimum equity investment of 20 percent, NCSC was able to obtain equity investments of 35–45 percent for the plant leases. "That gave us a significant equity cushion."

Between its incorporation in 1981 and 1986, when the Tax Reform Act severely restricted leasing opportunities, NCSC provided more than $1 billion in debt to facilitate CFC member participation in leveraged lease transactions, including four major power plant leases. NCSC's primary strategy was to try to help taxable cooperatives identify low-risk projects for lease transactions such as standard power plants using an established technology. "From the standpoint of equity investors, we tried to identify investors that had a perceived economic value in investing in the equity but also a familiarity with the technology and a familiarity with the industry," Lilly said. "The biggest player was General Electric." IBM and Shell were also important equity investors.

NCSC closed the first generating plant transaction for the lease of Colorado-Ute Electric Association's Craig Station Unit 3 on September 6, 1984. Craig 3 was a 408-megawatt coal-fired unit located in Moffat County, Colorado. Five equity investors purchased the unit for $522 million, including almost $300 million in equity. NCSC provided interim taxable debt of $228.1 million and arranged for a CFC guarantee of $102 million of tax-exempt pollution control bonds. The lease term was 26 years.

Colorado-Ute estimated that the lease would save $20 million per year compared to financing with an REA loan guarantee.

Leases Generate Savings

The lease of AECC's 35 percent ownership interest in Independence Unit 2, an 800-megawatt coal-fired plant located near

Newark, Arkansas, closed on December 6, 1984. A single investor purchased AECC's interest in the facility for $275 million with an equity investment of $126.4 million. NCSC provided the initial debt through an intermediate-term CFC loan of $148.6 million. AECC estimated that the plant lease would save its members $12 to $15 million annually over the term of the lease.

"What's interesting about this deal is that it was the first time a partial interest of a power plant had been leased," said Carl

National Cooperative Services Corporation helped CFC members save millions of dollars by participating in tax-advantaged financings, such as the leveraged lease for Arkansas Electric Cooperative Corporation's Lock and Dam 13 hydroelectric project.

Whillock, AECC's president and chief executive officer. "With CFC's help, we were able to sell the plant at its fair market value, which was significantly above the cost of construction. It was a tremendous thing, and CFC was involved from beginning to end."

NCSC assisted AECC in a safe harbor lease (a sale of tax credits under the Economic Recovery Tax Act of 1981) and a coal mine lease in addition to the Independence lease. "Through these leases we expected a savings of $23 million a year for our consumers," said Whillock. The G&T estimated that wholesale power costs were reduced by 12.5 percent, a savings of $70 to $75 a year for each consumer-member.

NCSC's third major plant lease was finally completed on December 7, 1984, although the formal leasing process started in late 1983 and was delayed until the commercial operation date of the generating plant. An equity investor purchased Seminole Electric Cooperative's Seminole Unit 1, a 600-megawatt coal-fired plant located in Putnam County, Florida, for $390 million, including an equity investment of $164 million. NCSC provided debt for the project in the amount of $225.9 million, including a $71.4 million variable-rate tax-exempt bond issue guaranteed by CFC.

Seminole estimated that its member systems would save at least $450 million over the term of the lease, compared to financing with an REA guarantee.

"Seminole was the most difficult project because of the pollution control bonds," Lilly said. "It took a year to complete. There were considerable lessons learned during that time that were useful in other transactions. The others took about six to nine months to wrap up, start to finish, and they all had to be concluded by a certain period of time because of the tax in-service dates. The deals had to be closed within a 90-day window from the time they were synchronized to the power grid. It put a lot of pressure on the cooperatives' staff, the CFC staff, the lawyers and the bankers to get the deals done, so we wouldn't lose the tax benefits."

The fourth major power plant lease was for Deseret G&T Cooperative's 96.25 percent interest in the 400-megawatt Bonanza Unit 1, a coal-fired facility. (The remaining 3.75 percent interest in the plant was owned by the Utah Municipal Power Agency.) The $684 million transaction included an equity investment of $306.1 million and NCSC debt of $377.9 million. The lease, which closed at year-end 1985, reduced Deseret's effective annual interest cost of financing the plant from 10.5 percent to 5 percent for a savings of $1.173 billion over the 25-year term of the lease.

"The Deseret leverage lease will stay in my mind for a long time because there was so much at stake," Crowson said. The loads on which the plant had been justified had failed to materialize, and Deseret had been unsuccessful in marketing the anticipated output of the plant. "We had to close by New Year's Eve of 1985 because of the tax laws. It was like a movie where people stayed up all night in the conference rooms drinking coffee and chain smoking, going from room to room agonizing. The last crisis of that day—I had been up all night—was that we had to get cash from the equity investor to pay off some tax-exempt bonds that were outstanding. To do that, we had to buy a particular kind of Treasury security by 11:00 a.m." There was just one problem: Crowson had no idea how to do that. "I remember walking hurriedly up the street to the Federal Reserve building with Marlene Fahey of Morgan, who mercifully did know what to do. I stood at a little shelf like you see in the post office and signed my name for the purchase of $146 million worth of bonds." The deal closed, barely beating the deadline.

CFC and National Cooperative Services Corporation worked together to obtain attractive financing for member systems for equipment such as this continuous miner.

One interesting project, which was not successful, was the attempt to lease North Carolina Electric Membership Corporation's interest in the Catawba Nuclear Station. "We made a very serious run at trying to do the first nuclear plant lease," Lilly said. "There were considerable hurdles to get over, including the major one of nuclear liability risk and who ultimately would be on the hook for that risk."

The leveraged leases provided tremendous benefits to the systems that were able to participate in them. "For co-ops that were taxable, it was a way of cutting the cost of a power plant by as much as 45 percent," Crowson said. "Arkansas Electric Cooperative Corporation is a very strong organization now, partly because of tax-driven deals. Colorado-Ute was saved for a while by the Craig 3 lease. Deseret has been helped by the Bonanza lease. Seminole has a small piece of Crystal River 3 and another large plant that is fairly expensive compared to the Florida Power and Light base. Without the lease I think they would have had a much tougher time."

In addition to reducing the cost for CFC's members, the leveraged lease transactions changed CFC's position from being a partial lender on a senior secured basis in a project with very little equity to the position of being, through NCSC, a 100 percent debt lender or guarantor with 35–45 percent equity in the project. Before leveraged leases, CFC had not financed more than

25 percent of the cost of a power plant. Although its level of committment increased, its risk decreased. "Not only did we lower the cost for the borrower, but we reduced our perceived credit risk by having an equity investment in those plants," Lilly said.

Safe Harbor Leases Generate Cash

Shortly after the creation of NCSC, the Economic Recovery Tax Act of 1981 created safe harbor leases, also called tax-benefit transfers (TBTs). The safe harbor leases provided a mechanism that allowed an investor in capital equipment to sell outright the tax depreciation rights associated with that equipment to another organization. The buyer gained the right to claim the tax benefits associated with the investment to minimize its federal tax burden. The seller received a cash payment for those rights. The concept was intended to stimulate economic recovery by encouraging organizations that could not use tax benefits to invest in capital improvements. A company that invested in plant and equipment could sell depreciation, including accelerated depreciation, investment tax credits and energy tax credits. "The transactions were actually invented to provide assistance to U.S. Steel and Chrysler," Crowson said. The law specifically said that organizations that were rate regulated on a return on rate-base method were not eligible to participate in safe harbor leasing. Cooperatives were not regulated in that manner, but most, not being taxable, did not qualify for tax benefits. Several systems became taxable in order to participate in safe harbor leases.

National Cooperative Services Corporation provided lease advisory services for safe harbor leases (tax-benefit transfers) for the following systems for a total asset value of $1.22 billion.

Sho-Me Power Corporation, Missouri

Western Farmers Electric Cooperative, Oklahoma

Associated Electric Cooperative, Missouri
(5 safe harbor leases)

Colorado-Ute Electric Association, Colorado
(2 safe harbor leases)

United Power Association, Minnesota
(2 safe harbor leases)

Arkansas Electric Cooperative Coporation, Arkansas

Seminole Electric Cooperative, Florida

Deseret G&T Cooperative, Utah

Western Fuels-Utah, Wyoming
(for the benefit of Deseret G&T Cooperative)

Allegheny Electric Cooperative, Pennsylvania

During the time that this option was available, NCSC arranged 16 TBTs on behalf of rural electric cooperatives. NCSC worked with the systems to identify potential projects for leasing, structure the transaction, place the equity and complete the transaction.

The TBT market was in many respects a purchaser's market. Those who wished to buy tax benefits demanded that the sellers indemnify them against the loss of anticipated tax credits. The sellers had to demonstrate that they could honor the indemnity. The corporations that were interested in purchasing tax benefits were generally unfamiliar with electric cooperatives, and they were only willing to work with corporations that had credit ratings. At that time, most of the cooperatives that had benefits to sell did not have credit ratings.

NCSC solved that problem by arranging credit support for the transactions with CFC. "Because those deals were put together so quickly and because of the uncertainty about interpretation of the law, the investment bankers persuaded us that a CFC guarantee was needed," Crowson said. CFC guaranteed the cooperative's indemnification to the purchaser of the benefit, which enhanced the member's ability to participate in safe harbor leasing.

Because of timing constraints, CFC held a special board meeting on April 26, 1982, to approve the first safe harbor leases for Associated Electric Cooperative and Western Farmers Electric Cooperative. "We flew into the Kansas City airport, went over to the Marriott Hotel, and spent several hours explaining what these were and why these particular credits should be accepted. Then we flew home," Bulman said. "We had a 90-day window after the project was declared commercial in which the tax-benefit transfer could take place. The law had recently been passed. It was a question of moving as fast as we could to understand what we could and couldn't do, and then do it."

The systems involved received a total of almost $500 million in cash payments. "CFC provided an immense service by getting that kind of money flowing into the industry," Crowson said.

Lilly agreed. "Considerable capital was transferred from the private capital markets to the cooperatives. That increased the system's equity level, as it received a large cash payment," he said. "The members did not have large tax obligations at the time. These tax benefits would otherwise have evaporated without any flow of capital to our members."

Nationally, more than $50 billion of TBTs were executed. The legislation was drafted in such a way that government entities such as states and municipalities could sell tax benefits in addition to private companies. "There were perceived significant abuses that caused Congress to rethink the legislation," Lilly said. Safe harbor leases were repealed by the Tax Equity and Fiscal Responsibility Act of 1982.

Variations of the leveraged lease program enabled NCSC to assist CFC members in other types of financings. When REA announced that it would no longer finance headquarters facilities for rural electric systems, NCSC developed a sale–lease-back program for depreciable real property, such as headquarters facilities, warehouses and other buildings. The program benefited distribution members as well as G&T systems.

The project financing program helped electric cooperatives develop generation from renewable resources, such as small hydroelectric projects. NCSC agreed to provide the debt that, along with an equity contribution from a project owner, financed construction of a generating project that would benefit a rural electric system. In 1984, for example, NCSC provided debt for construction of the 4.35-megawatt Cherokee Falls hydroelectric plant, located on the Broad River near Gaffney, South Carolina. Cherokee Falls Associates built and owned the plant. Broad River Electric Cooperative purchased the output.

Unfortunately, the Tax Reform Act of 1986 eliminated many of the tax incentives for investing in leased assets, including the investment tax credit and energy tax credits. It substantially curtailed leasing activities not only for electric cooperatives but for companies nationwide.

However, NCSC continued to be active in providing special financing services to the rural electric program as needs arose. In the 1990s, for example, it provided financing to for-profit affiliates of rural electric cooperatives for economic development programs.

CFC Supports Regulatory Reform

The high inflation and interest rates of the early 1980s along with other changes in the electric utility industry exerted financial pressures on CFC's member systems as well as on CFC. A decline in composite TIER and equity levels threatened member financial strength and CFC's ability to raise funds at attractive rates. "The financial condition of each member contributes to the cumulative results. To the extent that any member does not maintain its financial position, the economic health of the entire industry is weakened," Charles Gill said in the 1981 annual report. CFC targeted the systems that were not maintaining their ratios for special assistance.

One of the reasons systems were having problems was that rate increases were not being implemented fast enough to offset increases in costs. This was particularly a problem for regulated systems. CFC was frequently asked to provide witnesses in support of rate increases and to provide help in setting appropriate financial standards. It became apparent that someone was needed to do this on a full-time basis, and in December 1981, David Hedberg joined CFC to provide rate analysis and services.

"It was pretty clear that the regulated part of the industry was by far the financially weakest," Hedberg said. He adopted a goal of getting as many states deregulated as possible. "Cooperatives don't have the classic economic reason to be regulated, that is, the conflict between the stockholder and the ratepayer. Investor-owned utilities are always dealing with conflicts. They need to maximize the return to their stockholders, yet they have a public utility obligation. With cooperatives, there is a singularity of purpose. They should do what is in the long-term best interest of the ultimate consumer. It really makes very little sense for co-ops to be rate regulated."

In 1981, 25 states regulated the rates of electric cooperatives. "Most states were rate regulated simply because the cooperatives wanted to get some amount of territorial protection. They thought the rate regulation would be a nonconcern for them, because they did not anticipate asking for rate increases," Hedberg said. In the environment of the early 1980s, it was clear that rate regulation was very expensive and in some cases detrimental to the financial health of cooperatives. Total deregulation made sense in some states, whereas in others implementing a stream-lined approach, which provided an expedited procedure to bypass full review and commission hearings if certain criteria were met, was more practical.

The first two priorities were Virginia and Arkansas. In 1981, the average TIER for each of those two states was 1.20. "This was very, very poor, even by other regulated state standards," Hedberg said. CFC worked with the individual co-ops on a case-by-case basis, testifying in individual cases. Beyond that, CFC initiated an education program for the regulators. "We tried to educate them as to what cooperatives are and why they should be treated somewhat differently. That was something that very few states had done up to that point."

Work with Commission Staffs Pays Off

CFC's work with commission staff members was an important part of the strategy. "Most commission staffs are young, and there is a lot of turnover," Hedberg said. "Once they understood the nature of a cooperative, they were fairly open-minded to changing historical practices."

In some states, changing the cooperatives' attitude was more difficult. They appreciated the opportunity to shift the responsibility for rate increases to the commission. On the other hand, rate cases were very expensive. In the early 1980s, a typical case cost the cooperative between $25,000 and $50,000. By the early 1990s, the cost had risen to the $50,000 to $100,000 range. "I've been through several cases where even a distribution system spent several hundred thousand dollars. It's absurd," Hedberg said. Even the more reluctant cooperatives began to see the wisdom of improving the regulatory process.

Hedberg credited Arkansas attorney Leland Leatherman, a member of the Long-Range Study Committee and an incorporator of CFC, for much of the success in that state. When Hedberg talked to him about the need to reconsider regulation in Arkansas, Leatherman said he would arrange everything. "And he did," Hedberg said. "He had a lot of influence, and he fully supported our idea."

The turning point came when Hedberg was testifying before the Arkansas commission, and out of the blue the chairman asked, "Why are co-ops regulated in Arkansas?"

"He obviously saw it made no sense," Hedberg said. With Leatherman's help, they soon implemented a streamlined approach.

In Virginia, one of the commission staff members was particularly helpful. "Testifying in cases and talking with individual staff members was probably the area that was most successful," Hedberg said. "The first couple of times we met with resistance, because it wasn't the way that they had always done things. As we got to know people, they could see the point we were trying to make." Virginia, too, adopted a streamlined approach.

Soon cooperative systems in both states were much stronger financially.

CFC's efforts in this area earned the respect of the attorneys and engineering firms who offered services in this area, who might well have opposed it. "They generally took the approach that this would ultimately be in the best interest of their client," Hedberg said. Some, like Carl Stover of C. H. Guernsey & Company, began to seek out CFC's assistance. "He never hesitated to suggest to a client of his that they ask us to help on the financial side of it. We have very good relations with some of the consulting firms."

By 1994, only 16 states regulated the rates of electric co-ops. Seven of those had adopted a streamlined approach. Two others were very close to deregulation. This resulted in a significant improvement in the financial performance of systems in the deregulated states.

In addition to direct participation in rate cases and commission education, CFC offered rate seminars, which helped members understand the need for cost-of-service studies and cost-based rates. "We emphasized that we need to charge our members based on cost to the extent that's possible. That helps insure that we are getting the same kind of margin contribution from each customer class so we don't have problems with cross-subsidies," Hedberg said. "We got into a number of other rate areas, such as market-based rates, incentive rates and time-of-use rates. A lot of good things evolved from this process."

Over time, the rate group expanded its services into other areas, including evaluating acquisitions, property tax issues and other financial activities. "We have always tried to provide services that were in the financial area but that at the same time differentiated us from other lenders," Hedberg said. "To the extent that members have asked us to do a number of things that might not readily appear to be something a finance company would do, we've found that the member relations value and the specific savings we can contribute do a lot to differentiate us from other lenders."

The First Signs of Trouble

Several situations emerged during the 1980s that called attention to the credit risks in the rural electric program. Up to that time, electric cooperatives had an almost spotless loan repayment record. Since 1935, REA had foreclosed on only two systems and written off only $37,237. Both situations occurred in the 1940s. No supplemental lender had experienced a loan loss.

The first troubled situation to receive national attention resulted from the Washington Public Power Supply System's (WPPSS) termination of construction of nuclear Units 4 and 5. Thirty-five CFC members were among the 88 participants who had guaranteed payment of bonds issued to finance the project. The CFC members had guaranteed a total of 19.7 percent of WPPSS' financial obligations for Units 4 and 5. At the time of termination, the total investment in the project was $2.25 billion. WPPSS estimated that it would incur an additional $343 million in termination costs. WPPSS called on the project participants to pay their share of the projects expenses, even though they would receive no power or other benefits.

Numerous lawsuits were filed over various aspects of the WPPSS commitments, some of which were suits by WPPSS guarantors aimed at voiding their obligations. When WPPSS defaulted on $2.25 billion in bonds in 1982, it was the largest municipal bond default in history.

CFC itself had no financial relationship with WPPSS. However, it had advanced long-term loans of $68.1 million to WPPSS participants with additional commitments of $27.4 million. It had also advanced $5 million in line-of-credit funds.

The situation presented two important issues for CFC. First, it raised questions about the willingness of rural electrics to honor power contracts. CFC relied on power contracts between the G&T systems and their members as a basis for its G&T financing. Second, CFC had presented rural electric systems as people who always paid their debts. "The day the people in the Northwest started saying, 'We're not going to pay, even though we have a contract; we are going to find a technical way to get out of it,' CFC had a problem in terms of what we thought the marketplace expected of us and what we had led them to believe. That was the crucial issue," Bulman said.

CFC took the position that the systems should honor their WPPSS obligations. It continued to make funds available to members who were WPPSS participants, but it required them to pass a board resolution stating that the system intended to honor all obligations for termination and bond repayments of Units 4 and 5.

"If co-ops walked away from their obligations and we supported that, the whole economic underpinning at CFC would have been destroyed," Gill said. "We had no choice. What we suggested, I thought, was a legitimate solution." CFC offered to assist members in paying the WPPSS obligations into an escrow account until the issue of liability could be resolved. If they were relieved of the obligation, they could recover the funds. If not, they would be in a financial position to meet it. CFC even offered to make loans to cover the interim payments. In a further effort to be helpful, the CFC board allowed the WPPSS participants to deduct revenues generated to meet WPPSS debt service from revenues used to determine Subscription CTC obligations to CFC.

Some of the affected members strongly disagreed with CFC's position. They said that they had been misled when they agreed to participate in the project and that CFC should support their position. One system, Orcas Power & Light Company, even filed a voluntary petition for reorganization under Chapter 11 of the United States States Bankruptcy Code in an effort to repudiate the obligation, although it continued to make full debt service payments to REA and CFC.

"We were honest and straightforward. We expressed our opinion and showed people how to resolve the problem without harming themselves initially," Gill said. "I think we ultimately came across as a positive force."

Co-ops Excused from WPPSS Obligations

The Washington State Supreme Court and an Oregon state court found that the contracts were an extension of credit similar to a bond issue. The courts ruled that the municipalities in those states had no authority to guarantee the bonds because there had been no referendum for public approval of the obligation. Thus, the municipalities could not be required to pay WPPSS obligations. The Wyoming Public Utility Commission limited the ability of the Wyoming participants to raise rates to fund their WPPSS obligations. After those actions, a trial court ruled that the release of the municipal participants from their obligations so changed the circumstances that nonmunicipal participants must also be released. Orcas petitioned the bankruptcy court to withdraw its bankruptcy suit, and the petition was granted.

"We accepted the litigation that said they were not legally obligated," Bulman said. Making the payments could have subjected them to additional litigation from their members over use of co-op funds.

Although litigation on some aspects of the WPPSS situation continued well into the 1990s, these rulings essentially resolved the situation from CFC's perspective, and CFC returned to business as usual with its members. CFC experienced no loss and no lasting adverse effect as a result of its WPPSS experience.

The WPPSS situation was resolved relatively quickly. However, another situation was developing that would plague CFC and the rural electric program for years to come.

Wabash Valley Power Association (WVPA) was organized in the mid-1970s to provide power to 24 member systems located primarily in Indiana. Like many of the G&Ts organized at that time, WVPA pursued ownership of generating capacity through joint participation with other utilities. It entered into a complex series of negotiations with Public Service Company of Indiana (PSI), the previous power supplier for many of it members. In February 1978, WVPA agreed to purchase a 17 percent share of PSI's Marble Hill Nuclear Station, two 1,150-megawatt units. PSI retained the remaining 83 percent of the project. REA approved a loan guarantee in the amount of $466.7 million to finance WVPA's share. Like virtually all of the nuclear projects being constructed then, the unit experienced significant cost overruns. REA approved additional guarantees of $588 million for the project. In September 1982, the co-op acquired a 25 percent interest in Gibson 5, a 650-megawatt coal-fired power plant. REA provided a loan guarantee of $133 million for that project.

Crisis in Indiana

On January 3, 1984, PSI took unilateral action to terminate construction of the Marble Hill project on the basis that it was unable to raise capital to finance construction and that other resources were more cost effective. At that point, WVPA had borrowed approximately $466.7 million under the REA guarantees. When construction was terminated, REA told the system that it would no longer be able to draw additional guaranteed funds to make debt service payments on loans for the project.

As a result of these events, WVPA faced major financial difficulties. Debt service on the Marble Hill loan was $50 million per year. The cooperative could only pay that if it implemented an immediate 50 percent rate increase, an action that required the approval of the Indiana Public Service Commission (PSC; this name was later changed to the Indiana Utility Regulatory Commission). WVPA asked REA to advance $12.1 million to cover its March 1984 debt service payment. REA agreed to the advance, provided WVPA asked for a rate increase large enough to meet all of its requirements for the next 15 months and explored other options for meeting future debt service payments.

On April 6, 1984, WVPA asked the Indiana PSC to approve a 51.15 percent rate increase. It also began discussions with CFC about a loan to cover its debt service obligations until it could develop and implement a plan for meeting its obligations.

CFC and WVPA worked out a three-year phase-in plan that would get the system over the hurdles. In May 1984, the CFC board approved a loan to WVPA for $111 million to finance the phase-in. CFC attached several conditions to the loan, including receipt of a lien accommodation from REA, Indiana PSC approval of the rate schedules and a review of the member power contracts.

"We had a solution to the problem," Gill said. "We were going to lend them the money to transition over a three-year period. They were going to pay. The commission was on board. REA said no."

REA agreed to accept the phase-in only through April 1985, which would require borrowings of $53 million, but insisted that the full increase be implemented at that time. REA also rejected WVPA's request for a change in depreciation method. REA agreed to provide CFC with a Type II-A lien accommodation for the loan, subject to the necessary regulatory approvals. With the various time factors involved, this could not be accomplished before the July 2, 1984, due date for the debt service payment. CFC advanced $8 million plus funds to purchase Loan CTCs on an unsecured basis, based on the expectation that the Indiana PSC would approve the loan and on a letter from REA agreeing to provide a lien accommodation.

On September 28, 1984, the Indiana PSC authorized WVPA to borrow an additional $12.5 million to meet its debt service payment due October 1. It did not at that time approve the July borrowing. CFC refused to advance the October funds on an unsecured basis. WVPA decided that it would no longer try to borrow funds for the October or subsequent payments and defaulted. It asked the Indiana PSC to approve converting the July loan to a long-term secured loan, and on January 16, 1985, the Indiana PSC authorized that action.

In the meantime, REA took the position that by refusing to draw additional funds from CFC, WVPA had reneged on its agreement with REA and refused to provide security for the $8 million loan to CFC.

The relationship between REA and the cooperative deteriorated. WVPA said it could not request an immediate 40 percent rate increase (a portion of the necessary 51 percent increase had been implemented). It asked for help from the REA revolving fund. REA suggested a three-year phase-in. Noting that REA had recently rejected the approach, the cooperative said it was no longer viable. By early 1985, the word bankruptcy was being heard in WVPA discussions.

REA instructed WVPA to file for an emergency rate increase by February 1, 1985. On January 31, WVPA asked the Indiana PSC for permission to amend the $51 million request filed in April 1984 to seek emergency rate relief. On March 20, the Indiana PSC issued an order granting leave to file the amended petition but saying that no costs, including debt service, related to Marble Hill would be considered pending the outcome of an appeal in a case involving an Indiana PSC decision regarding another terminated nuclear unit, Bailly 1, owned by Northern Indiana Public Service Company. On December 27, 1984, the Indiana Court of Appeals had ruled that the Indiana PSC erred in allowing Northern Indiana Public Service Company to recover its Bailly expenses.

Federal Priority Statute Complicates Situation

Various efforts to develop a workable plan continued. Then REA played what it must have thought was its hole card. Title 31 of the United States Code, Section 3713, informally called the federal priority statute, is a 200-year-old law that says that, if a corporation defaults in payments owed to the federal government but makes payments due to others, the people responsible for making the payments can be held personally liable to the government for the amounts paid.

When CFC was first organized, the firm Milbank, Tweed, Hadley and McCloy, acting as general counsel, analyzed the impact of Section 3713 on CFC lending operations. Their conclusion was that the mortgage created an intercreditor agreement. Because the government entered the agreement consensually, the statute would not apply to proportional payments made on loans under the joint mortgage.

However, the government could try to apply it to the WVPA staff and board of directors who continued to make payments to other creditors, such as wholesale power suppliers. The government also denied the characterization of the mortgage as an intercreditor agreement and took the position that making debt service payments on pollution control bonds guaranteed by CFC, for example, while not making payments on the government's Marble Hill loans would expose WVPA staff and directors to personal liability. By a letter dated April 19, 1985, the Department of Justice notified the manager and directors of WVPA that they could have personal liabilities under Section 3713 if they continued to make payments to CFC or anyone other than the government while WVPA was in default to the government.

"It is fair to say that the program generally didn't know what Section 3713 was. Nobody had ever thought about it," said CFC Corporate Counsel John List. "It is also fair to say that the government, in an almost emotional reaction to the WVPA problem, issued the letter without realizing what the consequences would be. The government didn't adequately assess what the letter would mean to the individuals involved. The government saw it as a threat that would bring people to the bargaining table."

Instead, WVPA was advised that the statute did not apply to organizations in bankruptcy. After meeting with representatives of the cooperative, the Justice Department issued another letter on May 7 saying that the government would not attempt to recover normal operating expenses, excluding debt service to other lenders, paid through May 24 to give WVPA time to resolve the problem.

On May 23, 1985, WVPA filed a voluntary petition for reorganization under Chapter 11 of the Bankruptcy Code in the United States Bankruptcy Court for the Southern District of Indiana. It was the first major utility bankruptcy since the 1930s.

The CFC board was in session when a fax arrived announcing the co-op's decision. Gill and Bulman had reported that bankruptcy seemed likely. It was still a shock. "We were all literally stunned that someone would actually take an action like that as a utility, particularly one of ours," Sheldon Petersen said.

It signified a true loss of innocence for CFC and the rural electric program. Although there had been other difficult situations, up to that point, there was always an underlying assumption that things would eventually be worked out. As the WVPA case started to wind its way through the courts, it revealed weaknesses in the rural electric program and limitations in REA's ability to act that were previously not recognized. CFC found itself facing many changes in the basic tenets under which it had conducted its programs. That meant that CFC, too, had to change.

PART 5 CFC FACES A TRIAL BY FIRE

During the last half of the 1980s, CFC continued to expand the services and opportunities it offered its member systems and to provide leadership in addressing issues of national concern. It also focused attention on the need for cooperatives to expand their role in rural communities beyond that of electric service. CFC extended its programs to support nonelectric utility services and economic development programs. Interest rates declined throughout the period, and most systems were entering an extended era of stability. Unfortunately, a few troubled systems drew the attention of rating agencies and the financial community, and the creditworthiness of the entire rural electric industry was called into question. CFC devoted significant resources to assisting the troubled systems in their efforts to resolve their financial difficulties. Most of the situations were resolved favorably with little or no loss to CFC and no significant impact on CFC's ability to meet its members' needs.

Highlights

- *The National Rural Telecommunications Cooperative was incorporated in 1986 to provide telecommunications services in rural areas.*

- *In 1986, CFC established the Cooperative System Integrity Fund to assist cooperatives in fighting takeover attempts and defending themselves in territorial disputes.*

- *The Rural Telephone Finance Cooperative was incorporated in October 1987 to provide financing for rural telephone systems.*

- *CFC offered medium-term notes issued by dealers in 1988 and extended the program directly to member systems in 1989. This program funded specialized loans.*

- *CFC established the grantor trust program to help five systems prepay high-cost FFB debt in 1987 and 1988.*

- *CFC helped 20 systems buy out their REA notes in 1987.*

- *CFC's credit commitments, unadvanced loans and outstanding guarantees exceeded $10 billion for the first time in 1988.*

At the time WVPA filed for bankruptcy protection in May 1985, most of the rural electric program was entering an extended period of stability. Interest rates and inflation had declined substantially since the beginning of the decade. Major construction projects were drawing to a close. Many systems were turning their attention to rural economic development and other quality of life issues.

"Cooperatives were recognizing the fact that they weren't electric utilities. They were social organizations providing electric service. There were many other opportunities out there as well," said Sheldon Petersen, then an area representative. "Our job was to make sure that we were giving them the tools that they would need to fulfill their social purpose."

CFC's associate member program, established by the board in March 1983, slowly began to attract interest. CFC required the associate member to be organized on a nonprofit or cooperative basis and to be owned, operated or controlled by a CFC Class A, B or C member or by CFC itself. The associate member typically served the same constituency as its sponsor. Associate members were formed to provide such things as telecommunications services, water and wastewater services, and economic and business development opportunities.

The program gave cooperatives a way to maintain member loyalty by offering products and services in addition to electric service that would make them attractive to their members. "It also offered the co-ops the opportunity to grow load and spread their overhead. This could reduce their cost per kilowatt-hour," said Loan Officer Richard Bulman.

Pee Dee Electricom in Darlington, South Carolina, was one of CFC's first associate members. It was formed in late 1984 to lease, install and service satellite dishes for Pee Dee Electric Cooperative's members. "Most of our consumers couldn't afford to buy a satellite dish," said Robert J. Williams, Jr., manager of both co-ops. "We offer them an important service at a price they can afford. It also gives us the opportunity to visit with our members and to establish a personal link with them, something we haven't really had a chance to do in years. It's the best advertisement we've ever had." CFC provided a loan of $6 million to get the business started.

Guadalupe Valley Electric Cooperative formed an associate member, Guadalupe Valley Development Corporation, to serve as a catalyst for improving water services. Guadalupe Valley Development Corporation obtained a long-term loan from CFC, which it relent to water corporations serving Gonzales County and other areas surrounding San Antonio and Austin, Texas. The water corporations used the money to prepay Farmers Home Administration loans at a substantial discount.

"The financing that has been made available through our affiliated CFC associate member has enabled us to foster the improvement and stabilization of water services in the Gonzales area," said Doyle Hines, general manager of both systems. The improved water services helped foster economic development in the area. "Our involvement in assisting the water corporations has also increased the electric co-op's profile. In working with the water systems, we have come in contact with many community leaders that we may not have met otherwise."

Farmers' Electric Cooperative, Inc., in Chillicothe, Missouri, created Rural America Service Company to approach economic development from another perspective. Rural America Service Company borrowed $1.3 million through the associate member program to purchase a tract of land and to contribute toward water and wastewater systems for the Chillicothe Area Industrial Park. "If a rural electric cooperative in an economic area like ours is not looking at ways to support its members and to diversify itself, it has real troubles coming," said Farmers' General Manager Dan Bryan. "When we got ready to do the financing, we shopped around, because we wanted the best package we could find. I was extremely pleased when it turned out to be the one from CFC."

Not all of the associate members were successful. "There have been some notable failures, and those failures have been primarily because there was not enough commitment by the rural electric cooperative to support the associate member," Bulman said. CFC corrected the problem by requiring the electric system share the risk either by guaranteeing the loan to the associate member or by investing a substantial amount of its own capital in the project. "We forced the rural electrics to examine more closely the feasibility of these undertakings. In some cases, they decided themselves to back away from them, once they studied the economics more closely."

For systems with a solid idea and the expertise to develop it, the associate member program provided a valuable source of funds to improve rural life.

Bringing Telecommunications to Rural Areas

Driving the CFC programs in nonelectric areas was the need to make living in a rural area as attractive as living in an urban area. "In order for a cooperative to attract an industry to its area, it not only had to have an industrial park and a road system, it had to have good schools and a good quality of life. One aspect of quality of life that didn't seem to be available was television service," Bulman said.

CFC and NRECA jointly sponsored a study conducted by Southern Engineering Company of Atlanta, Georgia. The study analyzed the structure, technology, financing and future of telecommunications in rural America. "Telecommunications may well be as important to rural communities in the 21st century as electricity has been in the 20th century," said Gary Gordy, then CFC planning and member relations officer. "And rural electrics are positioned to play a key role in providing this technology to their members."

The study found economic feasibility for Television Reception Only services; bidirectional voice and data transmission between distribution systems, statewides, G&Ts and regional and national organizations; bidirectional video, voice and data transmission; and one-way video broadcasting—a national rural electric channel. The study said that fiber optics technology also had the potential to be attractive for rural electric communications in the future.

CFC and NRECA next appointed a National Rural Telecommunications Committee to implement the study's recommendations. Former CFC President James A. Vann, Jr., general manager of Dixie Electric Cooperative, agreed to chair the committee and its Management and Organization Subcommittee. Pee Dee's Robert Williams, Jr., chaired the Technology and Program Development Subcommittee, and Heywood C. Gay, executive vice president of Georgia Electric Membership Corporation, chaired the Member Relations and Marketing Subcommittee. The committee held four information and implementation sessions attended by more than 500 co-op representatives to present information about the study and a model program for systems interested in developing telecommunications capability.

Organizing NRTC

As a result of the participants' enthusiastic response, the committee was recast as the National Rural Telecommunications Council. It tackled issues such as an approved vendors list, technical standards and specifications, models for organizations, legal considerations, accounting procedures, CFC financing programs and marketing efforts. In May 1986, the CFC board accepted the council's recommendation to create a permanent

CFC supported the organization of the National Rural Telecommunications Cooperative to improve the quality of rural life by providing modern telecommunications services in rural areas.

telecommunications cooperative for rural electric systems by approving the organization of the National Rural Telecommunications Cooperative (NRTC). Approval by the NRECA board followed in June. NRTC was incorporated on August 6, 1986.

"The council has spent the last year developing an agenda and a sense of what this organization can do for both the rural consumer and the rural electric cooperative. With the approval, we can begin to put our first level of plans into effect and work toward a national, integrated telecommunications structure for rural America," Vann said.

CFC approved a loan of $75,000 to the new organization for

start-up expenses and continued to provide support services to NRTC until 1993, when the organization became totally self-sufficient. Bob Phillips, former counsel to the Kansas statewide, was appointed chief executive officer in 1987.

By 1994, NRTC had 800 telephone and electric members serving 48 states. It provided satellite and other advanced telecommunications and media services to its members and their affiliates. It also managed various business development projects affecting public and commercial telecommunications in rural areas. More than 80,000 satellite C-Band, or "home dish," customers were subscribers to its Rural TV service. NRTC negotiated programming contracts, provided a program authorization system and operations training and supported national marketing and consumer satellite equipment distribution by its members.

In 1994, NRTC entered into a joint marketing agreement with DIREC**TV**, a subsidiary of GM Hughes Electronics, that made NRTC the authorized distributor of DIREC**TV**, the first direct-broadcast satellite service, in many rural areas. NRTC expected to provide the service to more than five million homes nationwide.

"In addition to being able to sell dishes at a competitive price, NRTC's ability to negotiate and legislate programmers into offering programming at competitive prices was important," Bulman said. "It was a way to make the rural communities a little more attractive to industry, to keep people there who might have left, to give the co-ops a product that would make them useful and to give their customers a value other than electricity, particularly if their electricity was higher priced."

Financing Rural Telephone Systems

From the outset, the Long-Range Study Committee anticipated that there might be a need for CFC to provide financing to organizations providing rural telephone service. However, with widespread support from the entire telephone industry, rural telephone systems were able to pass legislation establishing a Rural Telephone Bank to provide supplemental financing. The Rural Telephone Bank was a quasi-governmental entity with lending criteria identical to REA's. That program was sufficient to meet their needs for many years.

By the fall of 1986, however, CFC was receiving regular requests for telephone financing. "There was a whole group of REA telephone borrowers that were not getting any value out of the concepts of CFC, such as jointly raising money. They needed to have access to an alternative source of financing, particularly in an era when the telecommunications industry was really beginning to be competitive," Bulman said. "More products and services were being offered. They needed to be able to raise money for things that REA traditionally did not lend for, such as long-distance networks, mergers and buyouts. It was a time when consolidations could improve the quality of service and sometimes lower the price."

The National Telephone Cooperative Association encouraged CFC to find a way to provide services to their members. CFC made several loans to NCSC to fund loans to telephone systems while the staff studied options for telecommunications lending. In December 1986, the CFC board approved the establishment of a new cooperative corporation to serve the specialized financing needs of the rural telecommunications industry.

The Rural Telephone Finance Cooperative (RTFC), a taxable cooperative, was incorporated in South Dakota on September 8, 1987. RTFC would be governed by an eleven-member board of directors, including six directors from CFC and five from the telephone members of RTFC. CFC President James H. Garner, a director of Randolph Electric Membership Corporation in North Carolina, was elected as the first president of the new organization. Membership in RTFC was open to systems eligible to receive REA loans to provide telephone services and their affiliates. NCSC's telephone loans were assigned to RTFC.

RTFC's relationship with CFC was similar to that of NCSC. CFC provided staff and management services as well as loans to fund RTFC's lending programs. In addition, CFC held a majority of the seats on RTFC's board and thus could consider RTFC an affiliate.

RTFC offered lines of credit and long-term loans of up to 15 years. Loans could be used for an array of capital needs, including projects not authorized by the Rural Electrification Act. Loan purposes included acquisitions, refinancing, central office equipment, networks, cable and wire facilities, support assets, cellular facilities and cable TV. Bridge loans provided immediate funds to systems with pending REA–Rural Telephone Bank loans.

In addition to loans, RTFC offered its members cash management services through CFC. RTFC members could invest in CFC commercial paper and use the paying agent service.

At the time, CFC had three major reasons for entering the telephone lending program beyond its basic desire to be of assistance to the telephone systems that, like the rural electrics, borrowed from REA. First of all, there had been discussion for some time about transferring the REA revolving fund to another lending organization. "We wanted to be positioned as a full-service organization that could absorb it all," said Petersen. "In the mid-1980s, CoBank [formerly Banks for Cooperatives] could lend to both sides, but CFC couldn't. We needed to enter the telephone business to position ourselves for the privatization of the fund, if that occurred." The revolving fund was subsequently dissolved by the Credit Reform Act of 1992 and with it that issue.

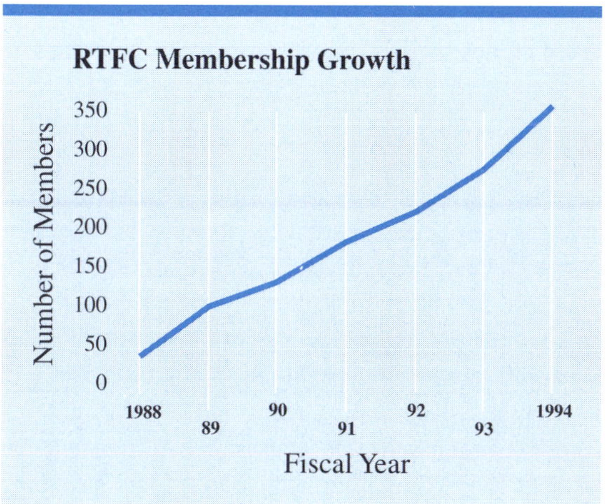

Since its inception, Rural Telephone Finance Cooperative has experienced significant growth in membership.

The other two reasons for supporting the telephone program continued to be important to CFC. First, it helped CFC build portfolio size and diversity. "These loans were similar to CFC loans, and they were big loans," Petersen said. That, in turn, helped to reduce the administrative and general expense component of CFC's interest rates.

CFC's involvement in the telecommunications industry offered a second advantage. "It is a place for us to learn about how to run a business in a competitive environment," Bulman said. "The electrics are going to be in a much more competitive environment in the next three to five years. There is a lesson to be learned from the telephone industry." The telephone industry already had competition for long-distance service and was beginning to experience competition for local services for some consumers. "We need to use that experience to say, 'Here's what you have to do on the electric side. Don't think that it is the same old business now.' We need to be willing to provide different options for electric service offering various terms and conditions."

One issue that had to be resolved was how the value of CFC would be shared with the telephone borrowers. Unlike the electric systems, the telephone systems were not required to invest in Subscription CTCs. Instead, CFC required RTFC borrowers to purchase larger amounts of the equivalent of Loan CTCs. "Over time, the telephone borrowers would hopefully get to the same equity level as the electric borrowers," Bulman said.

The leverage ratio of RTFC was projected to be about 7.5 to 1 compared with CFC's current ratio of about 4.5 to 1. Over time, changes in CFC's lending programs to balance equity investments with borrowings and provide faster retirement of patronage capital were expected to raise CFC's leverage ratio to about 8 to 1 as recommended by the Committee on Objectives and Planning.

RTFC

By the end of fiscal year 1994, the Rural Telephone Finance Cooperative had provided more than $1 billion in loans and loan commitments to rural telephone systems and their affiliates.

CFC experienced something of a culture shock in dealing with the telephone systems. There were about 1,300 telephone systems in the United States, including 900 that were REA borrowers. About 250 of those were cooperatives. The rest were commercial companies. "The biggest part of our telephone portfolio is actually lent to for-profit corporations," Petersen said. "The telephone program is a totally different program in that they don't have a passionate focus on REA financing. Their attitude is, if it's there, fine, if not, who cares. They have bigger problems. They have had to deal with an industry that has seen a technological revolution every three years. Rural electrics are still building the same wires and poles and transformers they did in 1935."

RTFC has been very successful in lending to telephone systems. By the end of fiscal year 1994, it had 353 member systems and outstanding loans and loan commitments totaling $1.1 billion.

CFC Assists New Co-op in Oregon

The citizens of Northeast Oregon provided CFC with a unique opportunity to contribute to the quality of rural life when they formed Oregon Trail Electric Consumers Cooperative in April 1987 to purchase the electric facilities of CP National Corporation. CP National, headquartered in San Francisco, had tried without success to sell distribution facilities serving about 35,000 consumers in Baker, Grant, Harney and Union counties to Idaho Power and Light Company.

"The proposed sale of CP National to Idaho Power concerned many people," said Peggi Timm, cofounder and president of Oregon Trail and treasurer of Baker County. "At one point CP National announced publicly its intent to release all of its employees. In a time when we were in the deepest economic depression of the decade, it didn't make sense to lay off all of the people who live here and replace them with workers from Idaho. I had a lot of friends who worked for CP National who were going to lose their jobs. I couldn't understand how that could improve our community."

Timm and other community leaders began to explore other

CFC commemorated the 50th anniversary of the Rural Electrification Administration in 1985 by presenting Pioneer Awards to those who had made outstanding contributions to CFC's organization and development.

J. K. Smith
Chairman of the Long-Range Study Committee, first president and first governor of CFC

David Hamil
REA administrator during CFC's inception and subsequent organization

Vincent P. Slatt
President of CFC's first elected board and member of the Long-Range Study Committee

Roger Oresman
First general counsel to CFC

Jerome S. Katzin
CFC's first financial advisor

Ira Shesser
Staff member to the Long-Range Study Committee and first loan officer of CFC

Charles U. Samenow
Consultant to the Long-Range Study Committee and first corporate counsel of CFC

Pioneer Award recipients recognized for contributions to the Long-Range Study Committee or as incorporators of CFC:

W. C. Carlton

J. R. Cobb

J. T. Dudley

Sailey Ennis

Walter Harrison

Virgil Herriott

Earl L. King

Leland Kitt

Leland Leatherman

E. V. Lewis

Owen Manning

R. B. Moore

Robert D. Partridge

Lloyd Patton

Clarence W. Peterson

options. They could try to find an investor-owned utility to purchase the facilities but did not find that option attractive. They considered forming a public utility district, but that required legislative action. Then they heard about electric cooperatives. "I felt the idea of a cooperative seemed to fit our lifestyle in this rural part of Oregon," Timm said. "A cooperative owned and controlled by the consumer-members it serves would bring control into the community and would encourage the community to work together to bring about positive economic change."

When CP National Corporation wanted to sell distribution facilities serving its Idaho community, Baker County Treasurer Peggi Timm enlisted the aid of other community leaders and CFC to form Oregon Trail Electric Consumers Cooperative in April 1987.

Less than four months after the termination of the Idaho Power talks, Timm and associates had formed Oregon Trail and opened negotiations with CP National. Neighboring Umatilla Electric Cooperative Association provided management and financial support to help the cooperative get started. "The question everyone kept asking us was, 'Where ya gonna get the money?'" Timm said. "A lot of people felt the cooperative would be a welcome change, but no one believed that the buyout of CP National by Oregon Trail could be financed."

Umatilla Manager Russ Dorran told them about CFC. David Hedberg and CFC Area Representative Harry Staley agreed to help the young co-op explore options. "The biggest concern we had was what kind of support there would be locally for the concept," Hedberg said. "If there was active support, virtually all things were possible. If they wanted us to do it all from Washington, it would never go anywhere."

CFC found much enthusiasm among the local business people, and CP National wanted to sell the property. CFC engaged an engineering consultant to appraise the condition of the system and to identify areas in need of improvements. CFC staff worked with Pacific Northwest Generating Company to construct feasibility models to determine the benefits to CP National consumers and to evaluate a fair purchase price.

The purchase was completed in October 1988. CFC provided a $33 million loan for the project and guaranteed a $12.5 million note to CP National. "We were extremely pleased with the financing," Timm said. "We were proud that our board was able to negotiate a purchase price $25 million less than the price discussed in the failed Idaho Power negotiations."

By 1994, Oregon Trail had gone from 100 percent debt financing to almost 20 percent equity while maintaining rates in the 5 cents per kilowatt-hour range. "Oregon Trail was more than just a buyout situation," Hedberg said. "It was a transaction reminiscent of the early days of the rural electric program. A struggling community, led by some dynamic individuals, successfully negotiated the buyout of an investor-owned utility for millions of dollars. I sensed great determination and caring in the people of northeast Oregon. This was a story with real obstacles to overcome and real heros."

New Financing Programs Diversify Funding Resources

CFC's greatest strength in going to the financial markets has always been the diversity of its portfolio. CFC continued to build on that diversity in the late 1980s by expanding its financial activities.

The CFC board approved a medium-term note (MTN) program in December 1987. Senior unsecured obligations with maturities ranging from 271 days to 30 years, the notes were designed to fund specialized loans such as equipment financing loans and RTFC fixed-rate loans. The MTN indenture was rated Aa3 by Moody's and AA- by Standard & Poor's.

The new program allowed CFC to minimize maturity and interest risks by match funding its loan assets with related MTN liabilities. Because the MTNs could be issued across the maturity spectrum, the average cost of funding was lowered. Because they could be offered on a continuous basis, they provided access to a broader range of potential investors.

Lehman Brothers and Merrill Lynch Capital Markets were chosen to market the MTNs. CFC's initial registration was for $400 million in MTNs. The first issue, $10 million in seven-year notes, took place in November 1988.

In March 1989, CFC extended the opportunity to invest in MTNs directly through CFC to its member systems. The member program offered investment opportunities for 10 to 24 months, or longer at CFC's option. The MTNs offered members a competitive rate and longer maturities than commercial paper.

CFC instituted several refinements in the commercial paper program, starting with overnight investments in June 1987 and expanding to intraday rates in June 1988. The intraday rates were adjusted by maturity and time zone throughout the course of the business day. "The change to intraday rates enhanced our ability to offer extremely competitive market rate returns on CFC commercial paper," Brandt said.

CFC Aids FFB Refinancing

CFC provided assistance to member systems that wished to prepay their higher cost FFB debt. Efforts to enable systems to refinance or repay the FFB debt without penalty began in 1986. "One critical factor in the cost of generating energy that has contributed to rate disparity is interest expense to our power supply systems," Charles Gill said. "G&T systems are currently being made to carry high-cost debt at a time when other electric utilities have been able to refinance a large amount of their similar debt."

Several bills enacted during the 1986–1988 period addressed the prepayment issue. Regulations issued by the Treasury Department and REA prohibited prepayments or strictly limited the number of systems eligible to prepay. Rules issued by REA in January 1987 allowed prepayments for systems experiencing financial difficulties. The legislation required the borrower to obtain the funds for prepayment from another lender.

CFC developed a specialized grantor trust financing program to assist members authorized by REA to refinance FFB debt. Soyland Power Cooperative became the first system to prepay FFB debt totaling $274 million in March 1987. CFC assisted four other cooperatives in refinancing FFB debt totaling $635.2 million in February 1988. With the exception of one other system that prepaid in 1988, REA did not approve additional prepayments until 1993, when regulations were adopted permitting prepayment under legislation approved in 1988. In 1991, CFC established the Guaranty Funding Cooperative

(GFC) to provide and service loans to CFC members to facilitate the refinancing of FFB debt. CFC transferred its refinancing variable-rate loans to GFC.

CFC Funds REA Note Buyouts

Although at 2 and 5 percent the interest rate on REA direct loans was not onerous, many rural electric systems wanted to prepay their REA debt to end their relationship with the agency. "The systems considered prepayment for purely business reasons, including the system's perception that REA had a greater cost than value, the fear that REA's interest rate would rise to market levels in the near future, concerns about REA's future and a desire to diversify in a nonpolitical environment," Gill said.

In fiscal year 1987, distribution systems were offered a one-time opportunity to prepay all of their REA debt at its discounted present value using the average yield on Aa-rated utility bonds as the discount rate. Twenty-nine systems took advantage of the opportunity, retiring debt totaling $687.1 million. The discounted prepayment amount was $406.1 million for a weighted average discount of 40.9 percent. CFC provided interim debt totaling $262.2 million to 20 systems and permanent financing through direct loans or guarantees of $295.5 million (including CTC investments). NCSC assisted one system in prepaying debt totaling $4.8 million at a discount of 36.1 percent, or $2.4 million.

Responding to Challenges to Cooperative System Integrity

For the first time in many years, the mid-1980s saw widespread challenges to the integrity of cooperative systems. In July 1985, the CFC board contributed $5,000 to the Electric Power Associations of Mississippi to assist Coast Electric Power Association in its efforts to resist a buyout attempt by Mississippi Power Company. NRECA also contributed $5,000 to the effort. The power company's unofficial offer was to buy out the equity, pay off the system's debt, reduce rates by about 10 percent and pay each member of Coast about $1,700 in cash. There were indications that the power company planned to make a similar proposal to Singing River Electric Power Association.

Shortly thereafter, there was an attempt by Iowa Southern

Utilities to buy out Chariton Valley Electric Cooperative, and the Iowa Association of Electric Cooperatives requested that CFC provide assistance in that case.

On August 30, 1985, CFC hosted a meeting of concerned members to discuss ways of resisting takeover attempts. About 40 representatives of systems in Mississippi, Iowa, Texas and other states attended the meeting. The group talked about strategies, ways to strengthen member loyalty and mistakes others had made.

It became clear that these were not isolated incidents and that a broader approach was needed. It was also another opportunity for CFC and NRECA to work closely together for the benefit of the entire rural electric program. At the September 1985 board meeting, the staff presented a proposal to develop a Cooperative System Integrity Fund to provide assistance to member systems resisting takeover attempts. A Cooperative System Integrity Committee, including representatives of CFC, NRECA and the Rural Electric Statewide Managers Association, would administer the fund, subject to the governor's approval, and serve as a crisis management team to provide assistance requested by the affected cooperative.

Bob Bergland, general manager of NRECA and a CFC board member, noted that NRECA had provided matching funds with CFC in the Coast and Chariton Valley cases. NRECA's board had authorized him to appropriate money as needed to assist systems, but a specific amount had not been set aside for that purpose. "NRECA and CFC need to deal with the economic problems that lead up to attempted takeovers and to assist the particular systems on an individual basis," he said. "Methods used in one state won't necessarily work in another."

Bergland also reported on a recent demographic profile of NRECA's membership. Almost two-thirds of the 11,000 rural electric system directors were active or retired farmers, but only 12 percent of the consumers served were active farmers. White-collar rural residents, who provided the greatest economic growth, were underrepresented on the systems' boards. "Any fund set up to assist member systems to combat takeover attempts cannot be used to help reelect directors when the membership is dissatisfied with the management of the co-op," he said in agreeing with comments by Governor Gill.

The CFC staff identified two possible mechanisms for

financing the fund, either through voluntary member contributions in the form of 5 percent of the next CFC patronage capital refund or by using CFC's Education Fund. The staff recommended against funding through nonvoluntary contributions, which would require an amendment to CFC's bylaws.

Of the 510 member systems responding to a CFC survey, 326 indicated a willingness to contribute 5 percent of their patronage capital to the fund. Most of the respondents agreed that establishing the fund was an appropriate rural electric objective. The CFC board formally approved the fund in January 1986. Voluntary contributions from the 1986 patronage capital refunds yielded almost $200,000 for the fund.

By the end of June 1994, the Cooperative System Integrity Fund approved grants of $3.5 million. Since 1986, the fund has helped 105 cooperatives fight takeover attempts and defend themselves in territorial disputes.

"The fund is here to defend against major takeover threats to service areas that require large amounts of cash to assist with court costs, legal fees and any other costs associated with protecting an organization's service boundaries," Petersen said. "When the co-op is stressed to the max and in need of financial assistance, the fund is there to provide assistance."

CFC Works to Resolve Troubled Situations

The greatest threat to the rural electric program during this period came not from hostile takeover attempts or municipal annexation efforts but from within. A few troubled systems became the focus for the rating agencies, the financial community and the courts. Although most of the rural electric systems did not share their problems, all were affected by the negative perceptions the difficult situations generated. CFC devoted significant financial resources, time and energy to assisting both member and nonmember systems in addressing their problems. In the end, both CFC and its member systems benefitted from that effort.

Although CFC performed an independent credit analysis for each of its loans, its role as a supplemental lender was such that, before the mid-1980s, it also relied heavily on its arrangement with REA in making loans. "In the early days, there were no loan requests that were so unusual or out of line that they raised serious questions as to whether we should make the loan," Ira

Shesser said. "We were quite comfortable with the conclusion that if REA was prepared to make a loan and they were providing a significantly larger amount of money than we were, it would be inappropriate for CFC to raise an objection." The vast majority of REA borrowers were in good financial condition. Those that weren't received 100 percent loans from REA.

This attitude continued as CFC developed more sophisticated loan and guarantee programs, increased the size of loans to individual borrowers and concentrated credit exposure in a few large G&T systems. "There was a feeling that if a loan went bad, the federal government would be able to solve the problem because of their implied powers under the Rural Electrification Act," List said.

During the 1980s, several G&T and distribution systems experienced financial difficulties severe enough to require action by REA and/or CFC. In most of the situations, the problem arose from the need for rates that were not competitive. A few situations involved attempts to abrogate the wholesale power contract between a G&T and its member systems rather than an immediate financial difficulty.

WVPA's financial difficulties were the direct result of the termination of construction on a nuclear power plant whose capacity, albeit expensive, could have been utilized by the G&T's member systems.

In the case of Sunflower Electric Cooperative in Hays, Kansas, and Deseret G&T Cooperative in Sandy, Utah, excess capacity that could neither be used by the member systems nor sold to another utility at a rate adequate to cover its costs elevated the needed rates to untenable levels. Soyland Power Cooperative had to deal with both the termination of a planned coal-fired unit and the absorption of its share of a nuclear plant that operated well but at a very high cost.

"If you are dealing with relatively small entities and they are making disproportionately large additions to their asset base, it creates a higher level of risk and much more uncertainty," Robert Edmiston said. "If the asset is not completed properly or on time, it introduces substantial financial risks to the system."

Regulation Complicates Troubled Situations

Regulation was a major factor in the troubled situations. Although only about a third of the G&Ts are rate regulated, nine of the eleven workouts for G&Ts occurred in regulated states. "When a regulator is there, it is almost impossible to solve the problem quickly and simply," Hedberg said. "If we have learned anything, it is that you have to get on top of a problem right away. The longer somebody doesn't pay you, the more difficult it is to deal with the problem."

Cooperative participation in nuclear power was a problem in some situations. Not all nuclear projects were troubled, not all troubled situations involved nuclear plants and not all troubled nuclear plants involving cooperative participants resulted in troubled cooperatives. Nevertheless, it was a significant factor that raised concerns about CFC's creditworthiness. "Nuclear exposure was an issue that affected the broad utility industry," Edmiston said. "Any company that had a broad nuclear exposure was viewed as having a greater credit risk." From the very beginning, CFC structured its operations and commitments in such a way that at no time would it be faced with a catastrophic demand for a large amount of money. The evergreen provision of the collateral trust bond indenture protected CFC from default if a member's performance fell. Under the guarantee programs, CFC would only be liable for annual payments in the event of a default by the guarantee recipient. In reestablishing its reserve for losses in 1984, CFC acted very early to provide further protection for its assets and position itself to maintain its financial performance regardless of what happened in the troubled situations.

In addition, effective June 1, 1985, CFC established a nonperforming and restructured loan pool to include any loans with principal or interest payments past due for 90 days or more, loans that were not expected to be repaid in accordance with the original terms because of a court order and any other loans when ultimate repayment was not expected. Members with loans in the nonperforming loan pool would not be eligible for patronage capital refunds until such time as CFC had been reimbursed for any losses or lower margins on the loans.

That done, CFC dedicated its staff and financial resources to addressing the problems in a positive way and protecting the integrity of the rural electric program. CFC worked closely with REA, the troubled systems—both members and nonmembers—the regulators and the financial community to create long-term solutions. CFC offered additional credit if that was what was needed to resolve the situation. Although there were some trying times, in the end CFC and the rural electric program came through them in good condition, and the lessons learned and changes in the program made it unlikely that similar situations would occur in the future.

WVPA Bankruptcy Changes Program Perceptions

Although WVPA was not the first rural electric system to experience financial difficulty, its May 1985 bankruptcy filing and the ensuing effort to restructure the system was precedential. "Wabash was really the wake-up call that things could go wrong in a big way," John List said. The bankruptcy filing followed REA's refusal to agree to a three-year rate phase-in plan, which would have allowed WVPA consumers to gradually absorb the impact of the cost of termination of construction of the Marble Hill Nuclear Station. CFC advanced $8 million in loan funds for the phase-in in anticipation of an REA lien accommodation for the loan.

The rating agencies and investment community expressed immediate concern about the bankruptcy and its implications for CFC and the rural electric program. "Things like that weren't supposed to happen," Brandt said. "One, it was a utility. Two, it was an REA borrower. The rating agencies said, 'I've always been told that's bankruptcy-proof.'"

CFC itself had believed it had little risk of a system not performing as long as REA was the major lender. "The Wabash bankruptcy showed how completely untrue that was," Hedberg said.

As the bankruptcy process proceeded, there were attempts to settle the various disputes, but none were successful. WVPA and its creditors entered a lengthy litigation process. The ensuing court decisions changed the way REA was viewed by CFC, the rural electric program and the investment community.

In April 1985, shortly before the bankruptcy filing, CFC had filed suit in the United States Claims Court in Washington, D. C., seeking to recover the $8 million advanced to WVPA in 1984. REA refused to provide the promised lien accommodation for the loan. In January 1988, the court ruled against CFC. Although a contract did exist between CFC and REA for the lien accommodation, the court said that not all the conditions had been met in order for the government to grant the lien accommodation. CFC appealed the decision to the United States Court of Appeals, which upheld the lower court. CFC charged the loan against the allowance for loan and guarantee losses.

"CFC made the $8 million advance with a sufficient belief that the funds were secured," List said. "We felt it was in the best interest of CFC and its members to assist Wabash." There had been other situations in which CFC advanced funds to help a member system in need while loan security was in process. As a result of the decisions, CFC became more diligent in formalizing REA approvals for CFC financings.

"We had done a lot of business on handshakes," Gill said. "As a result of the Wabash ruling, our attitude toward REA and handshakes changed."

The Indiana Utility Regulatory Commission did not rule on WVPA's April 6, 1984, request for a 51.15 percent rate increase until January 14, 1987, when the commission dismissed the case. The key issues were how WVPA should be regulated and whether the Indiana Supreme Court's ruling in a case involving another terminated nuclear plant was applicable to WVPA. CFC, REA, the Indiana Utility Consumer Counselor and some of WVPA's members were among the intervenors in the case.

Bailly Decision Sets Precedent

In 1982, the Indiana Utility Regulatory Commission approved a rate increase for Northern Indiana Public Service Company to allow it to recover the costs of its Bailly N-1 nuclear plant. The project was terminated before the plant was put into service. In allowing the costs to be recovered, the commission reasoned that, although the project was not completed, the decision to invest in it was prudent at the time it was made. On December 27, 1984, the Indiana Court of Appeals overturned the commission's ruling, saying that an investment had to be "used and useful" for a utility to be able to earn a return on it. On November 19, 1985, the Supreme Court of Indiana affirmed the Appeals Court decision.

Under Indiana law, rates for nonprofit utilities, such as cooperatives, were not based on providing a return on rate base. Instead, they were based on the utility's costs, including wages, materials and debt service. WVPA argued that the Northern Indiana Public Service Company case did not preclude the rate increase because WVPA's rates were determined by the Indiana law requiring the commission to set rates at a high enough level for WVPA to meet its debt service obligations.

The commission ruled that the recovery of costs associated with a canceled power plant was precluded by the Northern Indiana Public Service Company case, regardless of the rate methodology applied. "The cost of a utility plant never placed into service cannot be recovered through rates," the commission said, dismissing the rate proceeding.

In its intervention, CFC raised the question of who ultimately would be responsible for WVPA's financial obligations. The commission responded by saying, "We appear to reach a curious result, that no one is responsible for WVPA's debt." REA and CFC appealed the commission's decision, but it was upheld by the Indiana Appeals Court and the Indiana Supreme Court.

In 1985, subsequent to the events that precipitated the WVPA case, the Indiana legislature passed legislation that allowed recovery of investments in abandoned plants under some conditions.

REA Attempts to Set Rates

On November 23, 1988, after the appeals court decision, REA sent a letter to WVPA exercising its jurisdiction over WVPA's rates and directing the co-op to implement an immediate 9 percent rate increase, subject to the approval of the bankruptcy court. WVPA filed a suit in the United States District Court seeking a declaratory judgment that REA lacked the authority to preempt state law and exercise direct control over its rates. The district court granted WVPA's request in 1989, saying that, because REA had formal published policies deferring to state regulation, its letter was not adequate to assert control over WVPA's rates.

REA appealed the district court's decision to the United States Court of Appeals for the Seventh Circuit, which affirmed the lower court's decision. The appeals court said that whether or not REA had the statutory authority to preempt the state's regulation, it had not followed the rules of the federal Administration Procedure Act (APA) for doing so. "It sent Wabash a letter. There was no notice, no opportunity for comment, no statement of basis, no administrative record, no publication in the Federal Register—none of the elements of rulemaking under the APA," the court said.

The decision criticized REA's performance in several areas.

It pointed out that REA could have protected its interests in a number of other ways, including asking the state agencies to guarantee rates high enough to allow repayment of the REA loans, requiring the WVPA members to guarantee the loans or asking the members to authorize REA to set rates.

The appeals court also said that states are not always required to adopt rules of law that ensure federal lenders will recoup their investments. REA did not have special powers: in the case of WVPA, it was just another lender, with no rights or powers beyond that of any other creditor.

In April 1990, while the district court decision was on appeal, REA issued a notice of proposed rulemaking to consider substitution of federal regulation for state regulation of REA borrowers. After the notice and comment period, REA adopted two new preemption regulations, effective September 19, 1990. Essentially, the rules said that REA could preempt state regulation if the administrator determined that the state-approved rates were inadequate to allow the borrower to repay its loans and the borrower had failed to make required payments on secured loans. In addition, REA would have exclusive jurisdiction over the rates of any borrower in bankruptcy.

WVPA brought suit in the United States District Court to prevent REA from enforcing the rules, and the district court held the rules to be invalid. REA appealed the ruling. The United States Court of Appeals for the Seventh Circuit upheld the lower court's ruling.

In its decision, the appeals court said that the Rural Electrification Act did not establish a statutory basis for REA to assume regulatory authority over a cooperative's rates. REA, it said, must rely on its financing documents, including the wholesale power contract, the loan contract and the mortgage, to protect itself and ensure that loans are repaid. "Being a lending agency administering a federal credit program rather than a regulatory agency, the REA has at its disposal means to secure repayment that are less disruptive to the recognized prerogative of states to regulate cooperative utility rates," the court said. "As with any other lender, the REA assumes the business risk of advancing money to a specific organization, the risk that the organization will not be able to repay."

In July 1994, more than nine years after the initial petition for protection, the United States District Court affirmed an earlier bankruptcy court ruling in favor of WVPA's bankruptcy and reorganization plan and rejected the one that had been proposed by REA. Under the approved plan, REA would lose $486 million owed by WVPA. CFC, too, would experience an additional loss. The potential for that loss had long been factored into CFC's loan loss reserve. The ultimate resolution of the case appeared to be some years away, however, because of REA's decision to file an appeal.

CFC Grows from Experience

"The Wabash decisions changed completely how we look at credit risk, how we look at the role the government has in workouts and how we perceive what powers REA really has, especially when a cooperative is willing to fight REA and a regulatory commission stands between the co-op and the government," Hedberg said. "Our business is not isolated from credit risks that other industries have. We now take that into account when we are making credit decisions. We can't rely on REA to protect us. We have to protect ourselves."

List agreed. "Before Wabash, we thought that if there was trouble, we would stand shoulder to shoulder with the government, and we would solve the problem," he said. "Wabash showed us that we might have a very different view of how a troubled situation should be solved. We tried to solve the Wabash problem, but we were unsuccessful because we only represented 4 percent of the debt. When we couldn't get the government to come along, there was nothing we could do about it." CFC's experience caused it to take more aggressive action in other situations, such as in the Colorado-Ute bankruptcy proceeding and in Vermont, where CFC filed a foreclosure action.

It also highlighted a shortcoming of financing the cooperative form of organization. There is no provision for dealing with bad management decisions. "Normally, the people that are in senior management positions in a company or otherwise actively involved in decision making have equity positions in the company. They stand to lose a lot economically if they go into bankruptcy," List said. "In our program model, there is no direct individual economic loss to the members of the board or senior management, other than a possible loss of salary."

A regulatory agency can force the creditors to absorb a loss, but the result may be that creditors will no longer lend money to the system. REA, in fact, stopped all lending and lien accommodations for other lenders in Indiana, not just to WVPA and its member systems, but to all Indiana borrowers.

"REA was trying to exercise its leverage however it could to get Wabash to agree to a more favorable payment to the government than was proposed in Wabash's plan of reorganization," List said. The systems were able to finance additional construction out of internally generated funds for a number of years, but a devastating ice storm in 1991 forced the systems to seek additional loans. REA would not agree to lien accommodations, so CFC developed various unsecured loan programs that would make longer term loans available to Indiana members on an unsecured basis.

The court decisions changed the way REA operated. Historically, REA issued bulletins and sent out letters directing cooperatives to take various actions. "They had a pretty free hand in operating as a regulator/lender," List said. "Their view was that it was a consensual lending program of the federal government and that the requirements for notice, hearing, comment and published regulations did not apply." In the four years following the 1990 decision, REA published more than 700 pages of regulations in the *Code of Federal Regulations.*

REA subsequently insisted on a 20 percent equity level or member guarantees as a condition of G&T loans. That condition severely impeded the G&T lending program because of the complexity of putting the guarantees in place. "The theory is great, but like everything else, you can get lost in the details. It has become an unwieldy problem," Richard Bulman said. Should the guarantees be in proportion to load or number of systems? Is the guarantee proportional to the loan balance, or is it based on the entire loan amount? How are partial payments applied? What if another system doesn't pay? All of these questions would have to be answered before financing could proceed.

CFC agreed to act as agent for Alabama Electric Cooperative and its member systems in one complicated guarantee arrangement. Each of the co-op's 18 members had to file a mortgage document on the other 17 members. The final documentation for the loan was more than four feet tall.

"Once Wabash came along, it became a legalized society. You couldn't do handshakes any more," Gill said. "Wabash changed a lot of things. We changed our approach in some respects, but REA was also forced to change its approach."

Wholesale Power Contract Affirmed in Shoshone Case

One of the most significant situations to develop during this period did not involve a default in payment to either REA or CFC, but it threatened virtually every loan made to a G&T system by either organization. In October 1985, Shoshone River Power, Inc., entered into a Memorandum of Understanding with Pacific Power & Light Company (PP&L) to sell substantially all of its assets, including sales contracts, poles and power lines to PP&L. (PacifiCorp, a diversified enterprise headquartered in Portland, Oregon, conducts its electric business under the name Pacific Power & Light Company.) Shoshone's members subsequently approved the sale, and it took place in May 1986. PP&L next attempted to purchase the assets of Garland Light and Power Company, but the purchase was delayed by opposition from Tri-State G&T Association of Denver, Colorado, REA and CFC to the Shoshone sale.

Shoshone had been a member of Tri-State G&T Cooperative since 1958. It had entered into an all-requirements wholesale power contract with Tri-State that extended to December 31, 2020. The all-requirements contract defines the basic relationship between a G&T and its member systems. The G&T agrees to provide and the member system agrees to purchase all of the power and energy needed by the member system to supply its customers' needs at a rate sufficient to cover all of the G&T's costs. REA, CFC and other lenders have relied on this guaranteed stream of revenue as the basic security for the repayment of their loans. REA has routinely required systems to extend their wholesale power contracts to cover the term of any new loans. The contract enables G&Ts to borrow funds, even though their equity levels were relatively low.

Shoshone served a very attractive territory with a high load factor of almost 80 percent. Most of Shoshone's consumers were industrial customers involved in oil production. At the time, the large power loads stood to gain a rate advantage from switching to PP&L. "They were a major driving force in trying to get PacifiCorp to come in, and it was in their self-interest to do that," Hedberg said. "The irony is that Tri-State's rates have come down in almost a straight line since then, and they are very competitive today."

Shoshone informed Tri-State that, because it was selling its assets and ceasing its business operations, it would no longer have any power requirements and would no longer be purchasing power from Tri-State. Tri-State brought suit against Shoshone in the United States District Court in Wyoming to stop the sale and recover monetary damages. REA intervened in the case in behalf of Tri-State. Although Shoshone was not a member of CFC and CFC was not an intervenor, the issues involved were central to the operation of the entire rural electric program. CFC provided litigation support to Tri-State and REA in their efforts to enforce the contract.

Tri-State's position, and that of REA and CFC, was that Shoshone could not sell its assets without arranging for the purchase of electric power from Tri-State through the remaining term of the contract. The district court ruled in partial summary judgment that Shoshone was not required to remain in business throughout the term of the contract and that it would breach its contractual obligations only if it ceased operation in bad faith.

The remaining issues were determined by a jury trial held in April and May 1987. CFC Rate Director Hedberg testified in behalf of Tri-State and REA. "My purpose was to give the court a national perspective on the situation from the financial point of view," he said. "If PP&L could void the power contract by buying Shoshone, that meant that Tri-State and its members would have to pick up the fixed costs that were associated with the Shoshone load. Tri-State would have had to increase rates to compensate for the loss of that load. If additional co-ops were bought in the same way, the problem would escalate, and pretty soon Tri-State would have such high rates that they would cease to be financially viable."

Jury Upholds Power Contract

The jury found in Tri-State's favor on every issue. It said that Shoshone had breached its obligation to act in good faith, barred the purchase of Shoshone by PP&L and awarded Tri-State $37 million in damages. In June, the federal district judge disallowed all but $6 million of the damages and denied a permanent injunction to prevent PP&L from purchasing more cooperatives. The court ordered Tri-State to remit the remaining $31 million or face a new trial. When Tri-State refused to remit, the court set aside the jury's verdict and ordered a new trial on all issues.

Tri-State and REA appealed all of these actions to the United States Circuit Court of Appeals for the Tenth Circuit. The appeals court ruled that the district court erred in saying that Shoshone had no obligation to remain in business, noting several unique aspects of the all-requirements contract. It was a contract between interrelated parties, not an arm's length arrangement. In addition, REA was materially involved in the contract and the relationship between Tri-State and Shoshone.

CFC Director of Rates and Special Projects David J. Hedberg testified in behalf of Tri-State G&T Association and REA and provided other litigation support in an important court case that affirmed the G&T's ability to enforce the wholesale power contract with its member systems.

"The all-requirements contracts which form the Tri-State system are not simple requirements contracts but rather interdependent, joint and mutual contracts with a common purpose of securing the REA loans and thereby effectuating the REA policy to provide the economic means for supplying electricity to rural areas," the court said. The court further said that it was convinced that Tri-State and Shoshone expected Shoshone to remain in business and continue purchasing all of its power requirements from Tri-State throughout the term of the contract. "Indeed, the assurance of a long-term source of power and a long-term revenue stream is the very essence of the all-requirements contract and the Tri-State system; and the cooperative nature of the Tri-State system is dependent on the continuance of each member's contract," it stated.

Appeals Court Upholds Power Contract

The appeals court concluded that Shoshone did have an obligation to Tri-State under the wholesale power contract and that a new trial on that issue was unnecessary. It limited the new trial to the issue of damages. It vacated the order denying the permanent injunction against PP&L, saying that there could be no determination of that issue until the new trial determined whether future damages to Tri-State could be measured.

In other words, the appeals court upheld Tri-State's all-requirements contract with Shoshone. It was a very important ruling for the rural electric program. "When a distribution cooperative signs an agreement to purchase power from a G&T, that G&T must be able to count on the distribution cooperative being there for the life of the contract so that all debts can be paid," Hedberg said.

In November 1989, Tri-State and PP&L reached an out-of-court settlement on the damages issue. Tri-State received $22.5 million for power supplied to Shoshone since the May 1986 sale, plus interest, penalties and attorney fees. Shoshone remained a part of the PP&L system, but PP&L agreed to buy a significant amount of power from Tri-State until the 2020 contract expiration date to replace Shoshone's all-requirements contract. Garland remained a member of Tri-State. Tri-State valued the total settlement at about $100 million.

The Shoshone case highlighted three important legal issues. First, directors do have important fiduciary responsibilities. "Some of the people at Shoshone had not taken that as seriously as they should have," List said. Second, REA gained support for the principle that paying off distribution debt alone is not enough to secure a release of mortgage liens from the government. "They wouldn't release the properties unless a payment was also made in respect of the wholesale power contract," he continued. Finally, the court decision affirmed the integrity of the wholesale power contract. "The 10th Circuit came down very strongly on the side of defending the financial integrity of the program by saying that there is an implied covenant to remain in business for the systems that have signed wholesale power contracts."

The positive outcome of this litigation was very important to the integrity of CFC's lending program for G&T systems. "A number of investor-owned utilities were very interested in what was going on and what PacifiCorp was learning," Hedberg said. "They quickly saw what the ramifications could be if PacifiCorp could win the court case." The Shoshone result effectively ended the potential threat of that form of takeover at the national, as well as the local, level.

There were other challenges to the wholesale power contract during the 1980s. Coosa Valley Electric Cooperative stopped paying its power bill to Alabama Electric Cooperative, paying Alabama Power Company directly for Alabama Electric Cooperative's purchases instead. CFC and REA joined Alabama Electric Cooperative as plaintiffs in a case seeking to enforce the wholesale power contract and payment of Coosa Valley's power bills. The cooperative filed a bankruptcy suit in an effort to avoid its obligations under the contract. After several years of litigation over director elections, Coosa Valley emerged from bankruptcy through a settlement that left the wholesale power contract intact.

Southwestern Electric Cooperative, Inc., filed suit to terminate its contract with Soyland Power Cooperative. The United States District Court upheld the validity and enforceability of the contract in a countersuit brought by Soyland, REA and CFC.

Washington-St. Tammany Electric Cooperative, Inc., filed a bankruptcy suit in an attempt to void its wholesale power contract with Cajun Electric Power Cooperative, Inc. The reorganization plan included full repayment of all debt to CFC, REA and Cajun and a commitment by Washington-St. Tammany to honor its power contract.

"Before the problems that took place in the early and mid-1980s, there was a very strong presumption that members would honor their contracts, particularly in view of the fact that the distribution cooperatives created the G&Ts," Robert Edmiston said. The lesson of the Shoshone case and other challenges to the wholesale power contract was that the G&T, REA and CFC had considerable ability to enforce the contracts, even when the members did not want to honor them.

Cash Flow Crisis in Kansas Causes CFC's First Loss

The situation confronting Sunflower Electric Cooperative in Hays, Kansas, was a classic example of the excess capacity problem experienced by many G&T systems during the 1980s. Sunflower undertook construction of the Holcomb Station, a 280-megawatt coal-fired unit, in the late 1970s when load projections showed it would be needed in just a few years. By the time the unit was ready for commercial operation in 1983, loads were growing at a much slower pace, and Sunflower's member systems could not use the power. Because most of the area utilities were experiencing the same phenomenon, Sunflower was unable to market the excess capacity. A serious cash flow problem developed, because Sunflower was not collecting enough revenue through its sales to its member systems to cover its costs. Sunflower obtained financing from a group of banks to defer putting some of the Holcomb costs into its rate base, but by the summer of 1985, it was out of funds. The Sunflower board adopted a resolution asking CFC to help them.

On September 30, 1985, Sunflower failed to make its quarterly debt service payments to REA, CFC and other lenders. CFC had unsecured loans totaling $43 million outstanding to Sunflower at the time. CFC had approved the loans for front-end financing for Holcomb. Other creditors included REA, a group of banks that lent money for operating expenses under a revolving-credit agreement and a group of Japanese banks that issued a letter of credit to support Sunflower's pollution control bonds. The Witchita Bank for Cooperatives had also provided an indemnity guarantee in connection with a lease of certain assets. Sunflower had accumulated almost $650 million of debt on a system that was now valued at $400 million. The cooperative did not believe it could raise rates more than $5 million per year, and even that would require the approval of the Kansas Corporation Commission. Unless the debt could be restructured, Sunflower would have no choice but to ask for reorganization through the bankruptcy courts.

CFC helped Sunflower put together a debt restructuring plan that called for CFC to purchase the secured debt of the other lenders at a discount. CFC would receive first mortgage security for that debt and its unsecured loans, and Sunflower would request a $5 million rate increase. The CFC board approved the plan in October 1985. Under the proposal, CFC would have had a total of $405 million in secured loans to Sunflower, for which it would put up $362 million in additional cash.

The rating agencies reacted negatively to the proposal. Moody's put CFC under review almost immediately. "S&P did not put us on CreditWatch, but they were waiting to see if we intended to go through with the transaction," Truman Brandt said.

CFC was prepared to face the rating agency issue, but it could not find a way to resolve problems with the Kansas Corporation Commission. "We were going to do it on a variable-rate basis, because that was cheaper. The commission would not allow a flowthrough of interest rates," Gill said. CFC could not accept the interest rate risk, so the workout plan had to be abandoned.

However, CFC did not abandon Sunflower. It continued to meet with REA and the other creditors, along with the cooperative, to try to find another solution. Mindful of its experience with WVPA, REA through the Justice Department issued a waiver of liability under the federal priority statute, Title 31 of the United States Code, Section 3713, for the Sunflower staff and directors to allow additional time to develop a restructure plan.

CFC Covers Loss

The plan that evolved converted the debt into A, B and C notes. The A notes were given the highest priority for repayment, the C notes the lowest. Recognizing that it would likely suffer a loss whatever the outcome of the negotiations, CFC wrote off the estimated value of its C notes, $10.6 million, in 1986.

It was the first loss CFC had experienced, but it did not have a negative effect on CFC's relationship with the rating agencies and investors. "The market doesn't know if you have enough in your reserve until you have actually written something off and your reserve is proven adequate," Brandt said. When it is, they respond favorably.

Although the creditors agreed on the basic concepts for the debt restructure in August 1986, it took almost another year to document. The creditors signed the restructure agreement on June 30, 1987. It took even longer—until March 1988—to obtain the approval of the Kansas Corporation Commission. By that time, Sunflower had accumulated $662 million in debt.

The resolution of the Sunflower situation was possible because CFC agreed to provide a necessary future credit facility. CFC successfully negotiated a priority lien (ahead of REA) for any advances on the credit facility. The ability of CFC to meet future credit needs in workout situations with a priority lien became a key ingredient in numerous other workout situations.

As a result of its experience with Sunflower, CFC terminated its program for providing unsecured front-end financing, which had been a mainstay of its services to G&T systems since the mid-1970s. "We learned not to make unsecured loans for construction as a routine matter. It is a very risky thing to do," Bulman said. CFC subsequently required a lien accommodation from REA for construction loans, which REA only rarely agreed to provide.

Sunflower highlighted a risk other creditors faced in dealing with REA. "REA can plow all the debt they want into a rural electric system, and other creditors have no input," Bulman said. "There is no financial test in the mortgage that says the system must meet any TIER, equity or other standard before acquiring additional debt."

The Sunflower experience also affected CFC's efforts to develop credit resources in Japan. Although CFC was not directly involved in the relationship between Sunflower and the Japanese banks, the banks did not want to extend credit to any company involved in the rural electric program. CFC continued to call on them but made no progress. Finally, Brandt decided to go to Japan. He arranged to meet with the senior officials of the banks CFC hoped to develop a relationship with, and he addressed the Sunflower issue head-on.

"You may have some resentment that you took a loss in that situation," he told them. "Your problem was that you were dealing with a lead lender who did not understand how to make loans into an REA situation. If you really want to participate in this industry and to make money doing it, deal with someone who knows how to make loans to cooperatives. Deal with us."

The strategy worked. In time, CFC established credit with a number of the banks that had been involved in Sunflower and with other Japanese banks that had no previous knowledge of the rural electric program.

REA Buyout Solution for Illinois Valley

The situation confronting Illinois Valley Electric Cooperative (IVEC) was illustrative not so much of the problems facing troubled systems but of the difficulties facing all cooperatives trying to obtain financing from an increasingly complex REA. If much of the success of the rural electric program—and CFC—can be attributed to the support, effort and flexibility of REA, at least some of the difficulty during the 1980s can be attributed to the increasing number of constraints placed on REA's ability to act as a result of court decisions and administrative policy.

IVEC provides power to members in the area surrounding Princeton, Illinois, a distant suburb of Chicago. The factors that led IVEC to the workout table began to emerge in early 1980. IVEC was a member of Soyland Power Cooperative, a G&T whose rates inflated as a result of its participation in the Clinton Nuclear Station. Soyland was one of the first G&Ts to be allowed to refinance higher cost FFB debt as a part of a workout program. However, many of IVEC's problems were of its own making. After years of system neglect that resulted in a substandard quality of service and an irate membership, new management began a program to rebuild the system with $11.25 million in concurrent loans approved by CFC and REA. By 1985, the co-op had acquired almost $4,000 in debt for each consumer.

The parts of the system on which repairs had been completed were "gold plated," apparently without regard to a cost-benefit analysis, but much of the system remained in substandard condition. IVEC found itself cash short with rates 80 percent higher than neighboring Illinois Power Company and a membership that was no longer merely irate but actively demanding sellout. Although 13 utilities were notified of the potential sale, there were no acceptable offers. Unless CFC would agree to make line-of-credit funds available in the event of an ice storm, IVEC said, it would keep cash intended for debt service to REA and CFC on hand as a reserve for such a contingency. On December 1, 1986, the cooperative withheld the debt service payments totaling $435,428 due REA and CFC.

The First Workout

In January 1987, REA presented the IVEC manager and board president with a debt restructure plan that amortized all existing REA and CFC loans over a 40-year period, called for all new loans to be 100 percent REA loans at the 2 percent interest rate and permitted the system to operate at a 1.0 TIER on a cash accounting basis. REA estimated that the cooperative could then reduce its rates to within 30 percent of the Illinois Power rate. After some modification, the plan was implemented in November 1988.

"Unfortunately, the co-op management had no faith in the plan that was put together. When the cooperative does not believe a plan will work, it becomes a self-fulfilling prophecy," Hedberg said.

Beyond that, REA was unable to take timely action on IVEC's requests for additional loan support. By May 1992, the co-op was awaiting approval of an emergency loan application filed in October 1991 because of a severe storm that struck in March 1991, a lien accommodation to CFC for a line-of-credit application submitted in June 1991 and approval of a new two-year work plan submitted in April 1992. The two long-term loan applications totaled $5.7 million, and REA was also withholding unadvanced 2 percent loan funds totaling $600,000.

Even with rates 40 percent higher than those of Illinois Power, the co-op was adding load in the corner of its service territory that was within commuting distance of Chicago. "Prices for houses there are less expensive, because they are in the co-op service territory," Bulman said. "People are moving in there. They just don't use much electricity. They are managing their electricity usage and using other fuels for heat." The 1988 restructure plan assumed no load growth and $500,000 of annual plant additions. The actual growth over the next four years was 2 percent per year with necessary plant additions in excess of $1.5 million.

The Second Workout, an REA Buyout

In April 1992, the co-op informed CFC and REA that its board had authorized the payment of only 10 percent of its April debt service payment. "The workout was the government's workout," Bulman said. "The co-op had no ownership of the plan to make it work." Negotiations for a second restructure got underway. This time CFC adopted a different approach. "We said to Illinois Valley, 'You tell us the plan. If we are willing to accept it, we will take you on as a customer.'"

It appeared to the CFC negotiators that there were three

options. First, IVEC could file for bankruptcy protection. Aside from the fact the CFC would probably suffer a loss of $3 to $4 million, it could cost another $1 million in legal fees and take an extended period of time to resolve. The ensuing negative publicity would not be beneficial to CFC or its other members.

Second, REA could buy out CFC. This did not appear to be a promising approach, because CFC's debt was a relatively small portion of the cooperative's total amount of debt. Even if CFC agreed to write off all of its debt—an approach that was never considered—it wouldn't be enough to solve the problem. Aside from that, REA showed no interest in buying out CFC.

The approach that finally worked was for CFC to buy out REA's debt at a discounted rate of about 52 cents per dollar of debt. The restructure also allowed CFC to recover its B note of about $1.7 million from the previous workout, which it had not expected to recover, but it did increase CFC's debt exposure. REA relinquished all its interests in IVEC, but it prohibited the co-op from seeking or accepting additional REA funds in the future. The buyout reduced the co-op's debt from nearly $5,600 per consumer in 1993 to about $3,600 per consumer in 1994.

"The philosophy CFC took was to ask the co-op, 'Is this going to work? Can you live with these rates?'" Hedberg said. "We can do any kind of a study you want, but the bottom line is whether the management thinks it is going to work." Although the co-op would continue to have a high debt load and relatively high rates for several years, the commitment of its management and board offered a good chance that it would survive. The buyout took place on May 3, 1994.

The second IVEC workout was unusual in that it was the first time REA had agreed to a buyout in a workout situation and REA offered a significantly larger discount than the 10–12 percent in effect at that time under its general buyout rules.

"It is certainly a harbinger of things to come, not only in the remaining workouts but for other systems that are frustrated in dealing with REA," Bulman said. "Borrowers are less afraid to be away from REA than they were a few years ago. They can see that there is a life outside of REA, without REA second-guessing everything they do and having to wait for REA to approve rate changes, work plans and any new programs they want to try."

Colorado-Ute Bankruptcy Affects Entire Program

Colorado-Ute Electric Association was an anomaly among the workout situations. It was AA rated until the day it defaulted in its payments.

Colorado-Ute's financial difficulties became apparent when it announced that it would not make its first quarter debt service payments totaling $12.5 million to REA and CFC in 1989 due to liquidity problems. It also defaulted on obligations to other lenders. The system had experienced two consecutive years of losses. Its wholesale power rate to member systems of 3.9 cents per kilowatt-hour was one of the lowest in the nation, and it had not implemented a rate increase since 1981.

"Colorado-Ute's credit was essentially sound. What you had was an inappropriate rate structure that caused an unplanned, disastrous revenue shortfall, but the underlying properties and territory were economically sound," Edmiston said.

Both Moody's and Standard & Poor's downgraded the credit ratings assigned to Colorado-Ute after the announcement, but both rating agencies affirmed their ratings, Aa3 and AA-/A+, respectively, on bonds guaranteed by CFC.

CFC met with Colorado-Ute, its member systems, and other creditors and made several proposals for intermediate-term solutions. All were rejected by Colorado-Ute. CFC then developed terms for a standstill agreement that would have provided limited relief to all creditors until a long-term plan could be worked out. That was rejected. Colorado-Ute's proposals were for all creditors except REA and CFC to receive full interest payments during a standstill period or for CFC to provide an additional $40 million in loan funds to meet Colorado-Ute's debt service payments on existing CFC debt. REA and CFC could not agree to that.

On June 5, 1989, Manufacturer's Hanover of New York, United Bank of Denver and Norwest Bank of Minneapolis, which had a total of $30 million in unsecured loans to Colorado-Ute, petitioned the United States Bankruptcy Court in Denver to place the G&T under the court's control.

"Although we were willing to continue to negotiate in order to reach some middle ground, Colorado-Ute was not, and they walked out of the negotiations, virtually assuring the involuntary bankruptcy filing," Gill said.

At the time, Colorado-Ute was CFC's fourth largest borrower with $398.8 million in loans and guarantees outstanding. All but

$1.2 million was secured. Colorado-Ute had borrowed the $1.2 million just four days before it defaulted on its loan payments to CFC and REA. Its other outstanding debt included long-term securities, guaranteed financings and commercial paper, all of which had been rated.

"When you get a rating on short-term debt, you sign an agreement with the rating agency that you will keep them informed of any changes in your financial condition," Brandt said. "It states explicitly the things you will keep them informed about. Colorado-Ute didn't keep the rating agencies informed." As a result, some investors lost money on commercial paper, something that is never supposed to happen. "The problems should become known in sufficient time that commercial paper will have matured before a company's financial situation gets serious enough that it goes bankrupt," said Brandt.

As a result of the involuntary bankruptcy action against Colorado-Ute, Standard & Poor's placed CFC on its CreditWatch list with negative implications, meaning a downgrade was possible. Standard & Poor's cited concern over CFC's credit exposure to Colorado-Ute and another system with excess capacity, Deseret G&T Cooperative.[1]

Moody's also placed CFC's ratings under review for potential downgrade, except for the commercial paper program, citing the Colorado-Ute and Deseret situations.

Tri-State Offers Assistance

Several utilities offered assistance or buyout of all or part of Colorado-Ute's assets. Tri-State G&T Association offered a

---

[1] In 1981, CFC provided a guarantee for tax-exempt financing for pollution control facilities for Deseret's Bonanza I, a 400-megawatt coal-fired unit, when, for the first time, REA refused to approve a guarantee for the entire amount of the project. By the time the project was completed, anticipated load growth had not materialized, and the Deseret system had a significant amount of excess capacity. In 1985, CFC provided interim debt to NCSC for the debt portion of a leveraged lease of the project and later guaranteed NCSC's permanent debt for the lease. The lease reduced Deseret's effective annual interest cost of financing the plant from approximately 11 percent to approximately 5.8 percent. The substantial equity investment in the project served to enhance CFC's security in the Deseret system.

Deseret continued to be unable to market the output of the Bonanza unit at a rate sufficient to cover its costs. The cooperative and its major creditors, including CFC, entered into an Agreement Restructuring Obligations that reduced Deseret's debt service and rental obligations on the Bonanza plant until 1996, when large sales of power were expected to begin. The agreement closed in January 1991 with an effective date of January 1, 1989.

range of possible assistance measures from purchase of excess power to merger. Public Service of Colorado, an investor-owned utility that was among Colorado-Ute's creditors, and PP&L expressed interest in purchasing Colorado-Ute's assets.

CFC remained convinced that with internal cost restructuring, load growth, sales of excess capacity or assets and more realistic rate levels, it would be possible to solve Colorado-Ute's problems. "We are willing to consider any proposal that will serve the best interests of Colorado-Ute as well as the interests of our 1,000 rural electric members nationwide," Gill said at the time. "We are prepared to continue working with Colorado-Ute and its members and other creditors to reach a resolution."

On July 3, 1989, Colorado-Ute and its creditors were able to sign a five-month standstill agreement, which was later extended, designed to forestall the bankruptcy proceedings. CFC advanced $13.5 million to pay CFC and REA the principal and interest that was due in March, and Colorado-Ute agreed to pay interest on amounts owed to secured and unsecured creditors during the standstill period. Colorado-Ute also implemented an 8 percent rate increase. The banks that filed the action against Colorado-Ute supported the agreement, and the bankruptcy petition was dismissed.

The Colorado-Ute situation, along with other troubled situations, was beginning to have repercussions for others. The cooperative's lack of credibility raised questions about the entire rural electric program. "Colorado-Ute didn't hurt CFC directly from a credibility standpoint quite as much as it did the industry," Brandt said.

Downgrades for Other Systems

Citing difficulties with WVPA, Deseret and Cajun Electric Power Cooperative as well as Colorado-Ute, Moody's put eight G&T systems under review for possible downgrade on August 21, 1989. Approximately $1.8 billion of long-term debt was affected. "These situations underscore the need to reassess the financial and operating fundamentals of the G&T cooperative industry," the rating agency said.

In October, Moody's downgraded six of the systems, saying it was reemphasizing liquidity, cash flow and equity as important rating determinants. "G&Ts have more limited financial flexibility than investor-owned utilities to respond to the challenges facing the entire electric utility industry," Moody's said.

Many observers took exception to Moody's broad action. William Chew, senior vice president of Standard & Poor's, said that Colorado-Ute's difficulties were not symptomatic of the health of the rural electric industry. "Some co-ops will improve their credit strength in the near future, and others will get weaker. The industry, from a credit standpoint, will undergo a shakeout," he said.

CFC's senior staff stayed in close touch with the financial community throughout the difficult period during the late 1980s to emphasize that CFC and the rural electric program as a whole continued to be financially sound. Pictured, **left to right:** ***Richard B. Bulman, loan officer; Charles B. Gill, governor; Truman L. Brandt, chief financial officer; and John Jay List, corporate counsel.***

"It's obvious from the ratings opinions by Moody's that it has overlooked some important aspects of the cooperative power supply program and has superimposed the traditional tests generally applied in the analysis of the investor-owned utility industry," Gill said. "The G&Ts reviewed by Moody's are generally strong, viable, well-managed power supply systems." CFC's own credit ratings remained under review.

Colorado-Ute and its creditors continued to explore alternatives. In October, the Colorado-Ute board approved a letter of intent to merge with Tri-State and also voted to raise rates 14.5 percent, effective December 2. Tri-State agreed to lend Colorado-Ute up to $40 million under a revolving-credit agreement while the details of the merger were being negotiated.

Standard & Poor's placed Tri-State's A- rating on pollution control bonds on CreditWatch with negative implications after the announcement. Colorado-Ute's CCC rating on leveraged lease debt and C rating on commercial paper were placed on CreditWatch also, but with positive implications.

The adverse impact regulation can have on a workout situation was immediately apparent. Blaming the creditors for the G&T's troubles, the Colorado Office of Consumer Counsel opposed the rate increase. In response, the Public Utilities Commission rescinded the 8 percent increase initiated the previous summer, effective December 31. (Colorado-Ute could implement rate increases before commission review and approval.)

Colorado-Ute and Tri-State engaged Stone & Webster to perform a feasibility study on the proposed merger of the two systems. The results of the study, released in February 1990, said the merger would provide savings of $293 million through the year 2004 and free 200 megawatts of capacity for off-systems sales.

The systems continued to move toward the merger until April 1990 when the Colorado commission rescinded Colorado-Ute's 14.5 percent rate increase. Tri-State withdrew from the merger talks and suspended the cash advances it was making to meet Colorado-Ute's immediate financial needs. Colorado-Ute made a voluntary filing for bankruptcy protection on March 30, 1990.

Colorado-Ute Situation Precipitates CFC Downgrade

Moody's downgraded CFC's rating on collateral trust bonds from Aa2 to Aa3 shortly before the bankruptcy filing. On May 12, 1990, Standard & Poor's issued a downgrade on senior secured obligations from AA to A+ and on commercial paper from A-1+ to A-1. The actions were less a reflection of CFC's creditworthiness than a reflection of the downgrade of the entire industry.

The most important aspect was that business continued very much as usual after the downgrades. Series T was issued on May 23, 1990, shortly after the announcement of the credit downgrades. The $150-million seven-year issue had a coupon rate of 9.5 percent, below CFC's target of 9.75 percent, and was completely subscribed within a few hours after its release. The yield to investors was only 85 basis points over a comparable Treasury bond. Brandt estimated that the penalty from the downgrade was no more than 10 basis points. "We were very pleased with the investor reception to the issue," he said. "The nominal added cost, when blended with other available funds, did not have a noticeable effect on CFC's pricing of fixed-rate loans to its members. It was important that we quickly reassert CFC's traditional strengths in the bond market, and the Series T bond sale did that." There was no impact on the interest rate on CFC loans.

CFC Moves to Protect Interests

CFC applied one lesson learned from the WVPA experience immediately. CFC insisted that the bankruptcy court appoint a trustee to manage the system in behalf of the creditors. "Colorado-Ute had a series of managers who were very controlling and very limiting as to the information they provided to the board of directors and others," List said. "I don't think they would have managed the system in bankruptcy in a way that would have been open enough and responsive enough to the creditors' legitimate concerns." CFC led the successful effort for a court-appointed trustee to assume control of Colorado-Ute.

Victor Palmieri was appointed as the trustee in August 1990. Former chairman of the Overseas Development Council, he had been involved in the reorganization and liquidation of many large companies. "He had no baggage as a person from the program," List said. "He could step on anybody's foot that he wanted to, and he did. He didn't have to go to an annual meeting of the rural electric cooperatives and feel embarrassed about a position he took or worry about the next job he might get in the program. It helps explain why the reorganization was completed in two years, which was a pretty short time frame considering the magnitude and complexity of the problem."

CFC was among the creditors who initially questioned the selection of Palmieri as trustee because of his lack of cooperative experience. "Looking back on it, he was the right choice," List said. "He knew the issues. He marshalled a very good team. He was able to be very effective."

Palmieri was so effective that the consensual plan of reorganization filed September 26, 1991, allowed all creditors to recover their investments in Colorado-Ute. Under the reorganization plan, Colorado-Ute ceased to exist. It was the first workout that did not preserve the cooperative involved. Tri-State assumed the lease and became operator of the 408-megawatt Craig 3 station. It also acquired the 100-megawatt Nucla Station. Public Service Company of Colorado acquired 332 megawatts of Colorado-Ute's generating capacity and PacifiCorp 243 megawatts.

Ten of the 14 Colorado-Ute cooperatives became members of Tri-State. The remaining four elected to purchase power from Public Service of Colorado. "It was the first time that the membership had an individual decision as to whether they wanted to stay with the co-op or go," Gill said.

Rates were reduced for the former Colorado-Ute members to an average of 4.5 cents per kilowatt-hour, a 5 percent decrease from the rates in effect before the reorganization.

No Loss to CFC

CFC experienced no financial loss. All of its bankruptcy-related expenses were paid, and it was able to remove millions of dollars from its nonperforming loan category.

"This is a solution that benefits all the constituents of this plan, including the creditors and the rate payers, by maximizing the value of assets of the Colorado-Ute estate," Palmieri said.

Palmieri attributed the bankruptcy problems to a combination of economics, excess generating capacity and poor management practices. "If you recall the extent of the collapse of the western Colorado economy in the 1980s, it is not surprising that Colorado-Ute got into financial trouble," he said. "The key question was, once it got in trouble, why was it unable to deal effectively with its problems?" He laid the blame squarely on the shoulders of the former management and board of directors.

A cooperative facing a cash flow crisis has several options. "It can increase revenue by raising rates, if regulators will go along," Palmieri said. "It can decrease expenses by cutting administrative costs, it can renegotiate high-cost take-or-pay fuel contracts, it can reschedule debt service through negotiations with creditors or it can restructure both sides of its balance sheet by selling assets or merging with another utility. Colorado-Ute did none of these until it was too late. Why was that? Because the leadership failed to adjust its vision of the Colorado western slope energy future to the emerging realities. Instead, it simply tried to adjust the realities. It lowered rates instead of raising them in a vain attempt to build volume." The co-op also failed to maintain good relationships and credibility with the public service commission, its members, its creditors and the rating agencies.

An important lesson for other systems was that management, including the board of directors, has a fiduciary responsibility to the cooperative. In addition to the loss of the cooperative, the board members could have suffered personal financial losses. At one point, some of the creditors wanted to sue the directors personally for the money they were losing. "Those suits could have been prosecuted to the ultimate," Bulman said. "As a way of getting it behind us, we were able to convince people not to do that."

In the end, the best solution to a problem may not mean preserving a cooperative at any cost. "We would have preferred, obviously, that this bankruptcy and the developments leading to it never had occurred," Gill said. "But it did happen, and we at CFC felt it was necessary for us to play an active role in bringing about a solution that was in the best interest of CFC's members and the rural electric program. That is why CFC strongly supported the appointment of a trustee to stabilize the situation and to facilitate the reorganization. That is why CFC worked with the Colorado-Ute member systems and with Tri-State G&T to find an equitable solution. It speaks well of the electric cooperative industry that we were able to overcome adversity to keep our industry healthy."

Bulman attributed the positive final outcome in part to CFC's assertiveness. "I also believe Tri-State's role as a leading plan proponent helped tremendously," he said. In the end, the cooperation and dedication of the two organizations averted a catastrophe that could have had a negative impact on all rural electrics for many, many years.

A New Look at Risks

The troubled situations raised some serious issues that the rural electric program needed to address. "All of this was cumulative. The situations that came along raised three issues: one, the validity of the wholesale power contracts; two, the quality of management in the rural electric program; and three, the credit risks that are posed when a system is regulated by a state public service commission," Edmiston said.

Although the wholesale power contracts withstood some severe tests, the picture Paul Fox painted in the 1970s of co-op members as hard-working, honest people who always paid their debts was tarnished. However, by vigorously defending the contracts and insisting on enforcement, CFC was able to allay a substantial amount of the concern over this issue.

Some of the criticisms about the management quality in the rural electric program and whether the boards and staffs were properly equipped to make good management decisions over time were valid. Although throughout its history CFC offered programs to improve management understanding of financial issues and offered much direct management support, it was not an issue for which CFC alone could take full responsibility. “It is something that the individual systems have to address. It is something that is dealt with over time,” Edmiston said. “It is not an issue that is going to go away. The rural electric program is going to continue to be viewed by investors as having management that is, on average, not quite as strong as the management of investor-owned utilities. Moody’s has made a particular point of mentioning this in its credit rating review over the past several years.”

Many of the situations demonstrated that rate regulation by state public service commissions increased the risk in rural electric lending. CFC continued to respond to that concern very effectively through its rate services and by helping systems eliminate or streamline regulation in individual states.

CFC creditors continued to express concern about credit concentration. Investors do not like to see a large portion of a finance company’s loans and guarantee obligations resting with a few borrowers. “In 1984, when we received an upgrade, our top ten credits were 56 percent of the portfolio,” Brandt said. When CFC, along with the rest of the rural electric industry, was downgraded in 1990, the top ten credits had dropped to 40 percent of the loan and guarantee portfolio. CFC adopted a policy that addressed the issue by placing ceilings on line-of-credit and other loan amounts. By 1994, credit to the ten largest borrowers had been reduced to about 30 percent of the loan and guarantee portfolio.

“All of these events had an enormous impact on the confidence of investors in CFC’s obligations,” Edmiston said. “We spent a lot of time with investors and rating agencies, helping them to understand it and to assess the impact of these events on CFC’s creditworthiness.”

One of the ways CFC helped to allay investors' concerns was by responding directly and aggressively to the situations that arose. "CFC has resolved many of these situations very effectively," Edmiston said. "It has been absolutely tenacious in getting them cleaned up. CFC is a much stronger entity today."

Workout Principles

Over time, CFC established several principles for approaching workout situations. First, the cooperative itself had to be committed to trying to solve the problem. In most cases, that was true, but without the cooperation of the troubled system, there was little CFC could do to solve the problem. In addition, CFC learned to be more flexible. "We try to show whatever flexibility we can to help them solve their problems," David Hedberg said. "Rather than work as antagonists, we try to work as a partner with them to jointly find solutions that we can all live with. We have found some very good solutions in many of these situations, and we have come out of some bad situations very, very well."

CFC also established the principle that, in most cases, new money coming into a system should take precedence over old money. It began to make "super-secured" loans that would be paid before REA debt or other secured debt in the event of bankruptcy or another workout. This enabled CFC to help systems that had ongoing needs for loan funds despite their difficulties.

CFC also learned to say no. "We understand that we are here to provide the collective good for the program. Part of the process is to say 'no' to some individual systems," Hedberg said.

The heart of the process was standing by the members who had gotten into trouble but were doing their part to try to get out. "The message to the membership is that if you ever have the misfortune of being in this kind of condition, CFC will be there to help you work through it," Bulman said. "As long as you are straightforward with us, we will be straightforward with you."

Over time, however, CFC expects to be made whole. "There has to be a recognition on the part of the troubled system that the rest of the rural electrics that support CFC have to be compensated for any losses that a troubled system causes," he said. "They cannot be given special status. They can have a temporary special status while they work their way back to health, but they must make us whole."

CFC grew in many ways during this period. It recognized that some basic elements in the program were changing, and it responded by changing its programs and operations to reflect the new reality. This made CFC stronger, and it made the rural electric program overall stronger.

PART 6 CFC GETS READY FOR TOMORROW

The rural electric program entering the 1990s was different in several important aspects than it was when CFC was established. Largely as a result of CFC's efforts, the management and directors of cooperative systems were more sophisticated in their approaches to financial management. There was more diversity among the member cooperatives with fewer farmers and ranchers and more suburban, commercial and industrial customers. There was also less reason for members to be loyal to the cooperative form of power delivery. There had been fundamental changes in the REA loan programs and the level of service and support provided by REA. Although 25 years earlier electric cooperatives were unknown on Wall Street, there was now a high level of interest in lending to cooperative utilities. CFC responded to these changes by developing new programs, increasing the flexibility of programs already in place and becoming an even more member-oriented organization. CFC also provided leadership in the renaissance of rural economic development by structuring loan and guarantee programs to help member systems expand the ranges of services they could offer their communities. As CFC positioned itself to be the primary private lender and financial partner of the rural electric program, it prepared to lead its members into the next century.

Highlights

- *In August 1990, CFC initiated a three-year strategic planning process that resulted in several basic changes in CFC's programs and operations.*

- *CFC moved into new headquarters in Herndon, Virginia, in September 1990.*

- *The CFC board established the Member Advisory Committee in October 1990 to conduct a comprehensive analysis of CFC's policies, programs and resources.*

- *At the 1992 annual meeting, the CFC membership approved bylaws changes to elect directors at the district level and to allow the board of directors to determine the method of patronage capital rotation.*

- *Standard & Poor's upgraded its rating on CFC's collateral trust bonds to AA- in May 1992, returning all CFC credit ratings to the low AA level.*

- *In September 1992, CFC expanded its interest rate options to include mid-term fixed rates of 1 to 6 years and extended-term fixed rates for 8 to 30 years.*

- *The CFC board adopted a strategic plan in May 1993 that set priorities and outlined specific actions CFC could take to meet its goals.*

- *CFC changed its method of patronage capital retirement to the percentage method in 1993 and introduced numerous loan pricing initiatives to reduce its lending rate to borrowers.*

- *By January 1994, 19 systems had repriced their FFB notes with assistance from CFC.*

- *Standard & Poor's and Fitch Investors Service upgraded their rating on CFC's collateral trust bonds to mid-AA in May 1994, leaving only Moody's at a low AA.*

- *On May 25, 1994, Charles Gill announced that he would be retiring as governor of CFC at the 1995 annual meeting.*

- *CFC issued the first collateral bonds under a new indenture in September 1994.*

- *By November 1994, CFC had provided debt to nine members to buy out their REA debt (in addition to the systems that chose to do so in 1987).*

When CFC made its first loan in 1970, it was the only lender offering supplemental financing to electric cooperatives. Two decades later, competition was clearly the major issue, both for CFC and for its member systems. It became a dominant element in CFC's planning and operations. There was competition from traditional sources, such as CoBank and Wall Street investment firms, but there were also new competitors, such as G&T systems that developed funding programs for their member systems.

"We knew how to compete with CoBank, but we weren't exactly sure how we were supposed to compete with our own members," said Sheldon Petersen, assistant to the governor.

CFC's members faced their own competitive challenges. About 70 percent of the rural electric systems had rates higher than their municipal or investor-owned neighbors. Buyout efforts and annexation continued to be a problem.

The Energy Policy Act of 1992 introduced new levels of competition into the electric utility industry.

The legislation:

- Gave the Federal Energy Regulatory Commission authority to approve new exempt wholesale generators not subject to regulation under the Public Utilities Holding Company Act
- Amended the Federal Power Act to broaden the commission's authority to order transmission access

Some investor-owned utilities renewed their interest in acquiring cooperative service territory, possibly through buyout of entire G&T systems. The Energy Policy Act of 1992 created a new class of electric wholesale generators and opened transmission grids to competing utilities.

Most of the G&T construction projects undertaken in the late 1970s and early 1980s had been completed. In 1986, power supply construction work in progress represented 16.3 percent of total power supply plant. By the end of 1992, it had fallen to 2.8 percent of total power supply plant. Although there were individual exceptions, the program at large did not contemplate a new round of capacity additions for an extended period of time. With a 36 percent generation reserve, compared with 22 percent for investor-owned utilities, there was a significant level of excess capacity—and the excess cost that it entailed—but over time that capacity was expected to become an asset.

G&T debt costs remained high as most efforts to refinance FFB loans issued in the early 1980s were thwarted. G&Ts

found it difficult to obtain funds for new projects, such as transmission facilities, because of REA restrictions on lien accommodations and policy changes resulting from the litigation related to the WVPA bankruptcy and concerns about the integrity of the wholesale power contract. REA required G&T systems to maintain or demonstrate an ability to reach an equity level of 20 percent or to obtain loan guarantees from their member systems as a condition of further loans. This exacerbated tensions between the G&Ts and their member distribution systems. Throughout the program there was a high level of frustration with REA.

The question for the member systems was, how can relatively small utilities survive in an increasingly competitive environment? The question for CFC was, how can CFC help its members survive and be more competitive?

The G&T systems had always been unique. By 1990, the disparity among distribution systems had widened to the point that CFC could no longer rely on one-loan-fits-all programs. "In 1970, all the systems were basically alike," Charles Gill said. "Our product was basically the same for each of them. Co-ops had more invested here than they had borrowed here. It was very costly for them to go to another lender. Since then, cooperatives have become more sophisticated. We offer different things to different people now. There are certain cooperatives that use the same services now that they did in the 1970s. There are other systems that we can offer very little in terms of technical advice and assistance. They have the resources to do things for themselves."

It became clear to Gill that CFC needed to adapt its operations to the new environment. "CFC had come to a point where it could no longer rely totally on the loyalties of the membership," he said. "We came to the conclusion that CFC had to change if we were going to continue to grow and to do the job for the membership that they wanted. We had to change the bureaucratic nature of CFC. We had to change the systems, to change people, to change the whole organization."

CFC's Strategic Planning Process

Change does not come easily to anyone or any organization. The first step is knowing what to change. CFC began a strategic planning process in 1990 that continued over the next several years. It asked itself the questions, what does it mean to be a cooperative in the 1990s? What are our responsibilities to our members? How can we serve them best? What should we do to maintain member loyalty and support? The answers told CFC what to change.

In August 1990, the CFC board and key staff members met in Denver, Colorado, for a planning session. They talked about CFC's strengths, its weaknesses and future opportunities.

"The driving force behind the strategic effort was that we were having an identity crisis," Petersen said.

As it had in the past, CFC went to its membership for input. In October 1990, the board authorized the creation of the Member Advisory Committee. The committee was asked to conduct a comprehensive analysis of CFC's policies, programs and resources. The board also asked for guidance in three specific areas: CFC's role in enhancing the business environment and effectiveness of rural electric systems, CFC's role in future electric and nonelectric financing and CFC's role in providing financing products and services. It was the first broad member initiative since the Committee on Objectives and Planning had done its work eight years earlier.

Like its predecessors, the committee represented a cross section of CFC's membership, including two members from each of CFC's 11 districts and five members selected at large. The members included both managers and directors and representatives from small, medium and large distribution systems, as well as G&Ts, statewide organizations and national organizations. Bill McGinnis, general manager of Denton County Electric Cooperative in Denton, Texas, and a member of the CFC board, served as chairman of the committee.

Emphasis Shifts to Finance

Several things happened that shifted the emphasis in the committee's work from an introspective look at organizational and market issues to a much broader consideration of the entire financing system of the rural electric program. When the committee began its work, the authorized REA insured loan amounts were substantially less than the amounts authorized in the early 1980s. This created a backlog at REA of $833 million in insured loans waiting for approval on December 31, 1990, and a waiting period of 22 months for approval of a new loan. In addition,

The CFC Member Advisory Committee conducted a comprehensive review of CFC's policies, programs and procedures.

Bill McGinnis
Chairman
General Manager,
Denton County
Electric Cooperative
Denton, Texas

Ron Sheets
Chairman of Program and
Organization Subcommittee
President and Chief Executive
Officer, Kentucky Association of
Electric Cooperatives
Louisville, Kentucky

Billy E. Brown
Chairman of Capital Markets
and Financing Subcommittee
General Manager,
Withlacoochee River
Electric Cooperative, Inc.
Dade City, Florida

Richard K. Byrne
President and Chief Executive
Officer, Buckeye Power, Inc.
Columbus, Ohio

J. Malloy Chandler
General Manager,
Pioneer Electric Cooperative
Greenville, Alabama

O. B. Clark
President, Associated
Electric Cooperative, Inc.
Springfield, Missouri

Gary Coers
President, Shelby County
Rural Electric
Membership Corporation
Shelbyville, Indiana

Toxie Gerald
President, Cajun Electric
Power Cooperative, Inc.
Baton Rouge, Louisiana

Nadine Griffin
Director, D. S. & O. Rural
Electric Cooperative
Association, Inc.
Solomon, Kansas

Alton Higginbotham
General Manager, First
Electric Cooperative
Corporation
Jacksonville, Arkansas

Donald W. Kimball
Executive Vice President
and General Manager,
Arizona Electric
Power Cooperative, Inc.
Benson, Arizona

Paul J. Liess
General Manager,
Twin Valleys
Public Power District
Cambridge, Nebraska

Robert McClurg
President, Riverton Valley
Electric Association
Riverton, Wyoming

Robert L. McDuffie
Executive Vice President and General Manager, Randolph Electric Membership Corporation Asheboro, North Carolina

William C. McKamy
President, South Mississippi Electric Power Association Hattiesburg, Mississippi

Melvin D. Nicholas
General Manager, Eastern Iowa Light & Power Cooperative Wilton, Iowa

Lynn Puckett
General Manager, Red River Valley Rural Electric Association Marietta, Oklahoma

Randall Pugh
President and Chief Executive Officer, Jackson Electric Membership Corporation Jefferson, Georgia

ODeen K. Redman
Director, Raft River Rural Electric Cooperative Inc. Malta, Idaho

B. Dean Sanger
Chief Executive Officer, KAMO Electric Cooperative Vinita, Oklahoma

Eugene P. Sullivan
General Manager, Stearns Cooperative Electric Association Melrose, Minnesota

I. Wayne Swann
President and General Manager, Southern Maryland Electric Cooperative, Inc. Hughesville, Maryland

Richard J. Thronson
General Manager, Valley Electric Cooperative, Inc. Glasgow, Montana

David E. Turner
President, Warren Electric Cooperative, Inc. Youngsville, Pennsylvania

Henry Umscheid
General Manager, Bluebonnet Electric Cooperative, Inc. Giddings, Texas

Bert Voegele
General Manager, Codington-Clark Electric Cooperative Watertown, South Dakota

Wayne Willink
President, Dairyland Power Cooperative La Crosse, Wisconsin

William W. Wood
General Manager and Chief Executive Officer, San Isabel Electric Association Pueblo, Colorado

responding to pressures over the budget deficit, Congress reduced the fiscal year 1991 authorizations by 25 percent with a provision for a 90 percent REA guarantee of private sector loans to offset the reduction. The guarantee provision could not be implemented because REA did not publish the rules that would govern the guarantee until 30 days before the end of the fiscal year. As implemented, the 1991 authorization provided $485 million for insured, concurrent electric loans.

REA proposed substantial changes in the financing program for fiscal year 1992 with an increased reliance on loan guarantees and the use of a means test to allocate available insured loan funds.

In response to these issues, NRECA formed the Cooperative Finance Study Committee to assess the financing needs of the rural electric program, particularly as they related to federal financing through REA. In addition, the G&T Managers Association created a Future Financing Committee to evaluate G&T alternatives.

The CFC Member Advisory Committee and the NRECA committee held a joint meeting on May 1, 1991, to discuss and coordinate their respective activities. One of the issues on the agenda was the work of a group of cooperative managers, the research committee of the Rural Electric Management Development Council. They had recently published a study proposing the creation of a bank for utility cooperatives within the framework of the FCS by transferring the assets and the liabilities of the REA revolving fund to the new bank.

The NRECA committee asked the Member Advisory Committee to prepare an alternative plan based on the model offered by Fannie Mae. Fannie Mae is a stock corporation whose function, among other things, is to purchase home mortgages from retail lenders and remarket them to investors. Congress created Fannie Mae to make capital available for home mortgages and provided the initial equity. Subsequently, Fannie Mae became a privately held corporation.

Fannie Mae is similar to CFC in that an investor can buy into a specific pool of mortgages. However, with Fannie Mae, the revenue stream from the mortgages is paid directly to the investor, whereas CFC acquires mortgages on utility properties and pledges the mortgages as collateral for its bonds. Several statewide associations also asked the committees to give serious consideration to the Fannie Mae concept.

One possibility was for the government to transfer the assets and liabilities of the revolving fund to an organization such as CFC. That would allow the rural electric program to control directly the operation of its financing source. Because mortgage-backed securities enjoyed wide market acceptance, the committees thought that this might be a more appropriate financing vehicle than a government-sponsored enterprise (GSE) like the FCS. "The whole concept, which came to be known as privatization, was an attempt for the rural electric systems to gain control of their own financing at REA," Gill said.

Privatization Dominates Discussions

Once raised, the issue dominated the rural electric industry. "The minute that issue hit, almost everything else that the Member Advisory Committee was dealing with took a back seat. That was a major, major program event," Petersen said. "We still dealt with things like director elections, but privatization became the thrust of the committee's work."

Legislation in the form of the Federal Credit Reform Act of 1990 heightened the need for quick action if the program decided to pursue transfer of ownership. The act created a new financing account as of October 1, 1991, that would make all future REA insured loans on a "pay-as-you-go" basis. The new account would initially have a difference between assets and liabilities of about $2 billion in equity. In the absence of appropriations, however, all defaults, including G&T defaults, would be charged against the account. That could deplete the equity in the account in as little as three to five years. Thus, if no action were taken, the program could lose a valuable asset that could be used to reduce interest costs well into the future.

In July 1991, the Member Advisory Committee presented a plan to the NRECA committee. The CFC committee concluded that it would be economically feasible to transfer the approximately $12 billion of direct loans from the REA revolving fund to a private organization such as CFC or a GSE similar to Fannie Mae. The Member Advisory Committee assumed that if a GSE were created, CFC's assets and liabilities would also be transferred to it, but there was a significant question as to whether the current political climate would allow the creation of a new GSE.

"The amount of money owed to the Treasury was less than the assets of the revolving fund, so there was equity in the revolving fund," Gill said. "On a present-value basis, it had very little value to the government." The entity receiving the revolving fund could continue the same schedule of payments to the Treasury and still have the remaining funds available for new loans.

The committee estimated that the fund could provide approximately $600 million in annual loan funds at a perpetual lending

The Member Advisory Committee identified six key objectives for CFC.

- To provide rural electric distribution systems with an assured source of capital at the lowest possible cost at whatever level is required
- To provide rural electric power supply systems an assured source of capital to supplement or enhance funds provided through federal guarantees
- To establish credit fundamentals for CFC and the rural electric program that will assure bond ratings for CFC securities at or above a mid-AA level
- To provide resources, as required, to meet the evolving needs of the rural electric program for capital and financial services
- To operate and maintain CFC in a manner that will foster and develop cooperative principles of organization and philosophy
- To foster and support the maintenance and strengthening of the common goals and unity of purpose of the rural electric program

rate of 5 to 7 percent. The committee said that any organization receiving the fund should be controlled by the cooperatives. It left to NRECA the task of determining whether such a plan would be politically acceptable to the membership, Congress and the administration. If the plan were not politically acceptable, the committee said, then there should be legislation that would allow individual systems to prepay REA loans at a favorable discount rate.

The committee also said that G&Ts should be able to refinance FFB debt without the economic hardship of the existing premiums and penalties and that REA should continue to provide guarantees for power supply financing.

Options for Future Financing

Two options for future financing were discussed at the fall 1991 regional meetings: the transfer of ownership plan and the continuation of the REA program. "Our political strength will maintain REA for some of us, I am sure, for the foreseeable future," Gill told delegates to the meetings. "However, can we sustain it for all, or as importantly, can we convince those systems that have a strong desire to remove themselves from the program to stay with us over the long haul? Our economic and political strength is based on each and every one of us working together. We have said for many years that REA is the glue that holds us together. In light of recent actions by groups within the program, perhaps a below-market pool of capital, rather than REA, is what holds our program together."

The Member Advisory Committee recommended that CFC amend its bylaws to allow director elections to be held at the district level.

Although financing issues dominated the discussion at the regional meetings, Member Advisory Committee recommendations in several other areas were also presented to the membership. One related to director elections. "One of the matters that has, over the years, caused as much conversation as any other matter involving CFC, is our method of electing CFC directors," Gill told meeting participants. "The Member Advisory Committee thoroughly investigated this matter and reviewed a number of different options to our current practice."

CFC's bylaws provided for the directors to be selected through a series of district and national elections. The concept was that, because the directors represented the entire membership, the entire membership should participate in electing them. Although good in theory, the practice was not satisfactory for numerous reasons, including difficulty in getting candidates to oppose a nominee who was considered the front-runner. The Member Advisory Committee recommended a procedure that would allow elections to be held at the district level and eliminate many of the concerns about the original election procedures.

The change would assure that each district would have the representation that it wanted and eliminate expensive national campaigns. The committee recommended that the bylaws also be amended to eliminate any age limitations for board service.

Fine-Tuning CFC's Capital Structure

The Member Advisory Committee addressed the need to make two changes in CFC's capital structure. "CFC must be certain it can provide the private-market capital needed, in whatever amounts, and we must be able to maintain a double-A credit rating to be of maximum benefit to the members," Gill said at the meetings. "The double-A rating is important, not only from the cost standpoint, but also in terms of the availability of capital."

At that time, CFC had about $4.50 of debt and guarantees for every $1.00 of equity for a debt-to-equity ratio of 4.5 to 1. "We believe that in the long term CFC must keep its debt-to-equity ratio in the range of 6.5 to 1 to 8 to 1," Gill said. The higher ratio would allow CFC to increase the amount of capital it could raise. The rating agencies were continuing to express concern about the amount of capital available to CFC to absorb potential operating losses. All of CFC's equity capital, CTCs, other subordinated members' debt and allocated patronage capital was committed to the members. "The issue that this raises to investors is that if CFC experiences a loss greater than its reserve and current year's margins, there is no balance sheet equity to apply it to and it remains on CFC's balance sheet as a loss in perpetuity," Gill said. "To a degree that is as much cosmetic as it is substantive, but it does affect how Wall Street analysts view CFC. They are accustomed to viewing the balance sheets of for-profit corporations and their experience affects their understanding and acceptance of CFC's credit."

The Member Advisory Committee made several recommendations related to CFC's financial base. First, CFC should increase the capital available for potential losses by allocating patronage capital at the time of its retirement rather than at the close of the CFC fiscal year. CFC could tell members each year how much patronage capital they had earned, but they would probably not be able to include it in their income statements until it was received.

To stabilize CFC's debt-to-equity ratio, the committee

recommended that CFC extend its patronage capital rotation cycle up to 10 years. The committee said that this would enable CFC to continue to meet the financial market requirements for a double-A credit rating while allowing the loan portfolio to continue to grow. The Member Advisory Committee also recommended that the bylaws be amended to allow the board to determine the method of retiring patronage capital. (The original bylaws required patronage capital to be rotated on a first-in, first-out basis.)

CFC Governor Charles B. Gill supported the Member Advisory Committee's recommendation that the requirements for membership in CFC be further liberalized to encourage the few remaining nonmembers to join.

Finally, the committee said that CFC should establish TIER goals in the range consistent with maintaining bond ratings in the mid-AA range. Although this had always been an informal goal, it was the first time CFC contemplated a formal policy.

Despite significant efforts by the staff and board of directors during Gill's tenure as governor to encourage the remaining nonmember systems to join CFC, about 90 electric cooperatives still were not CFC members. CFC had offered several options for meeting the Subscription CTC requirements, but many nonmembers were deterred by the requirement for significant up-front equity investments. The cost tended to make CFC uncompetitive on the first loan on a strictly financial basis. The Member Advisory Committee recommended that membership requirements be further liberalized by allowing distribution and power supply systems to become members by paying the $1,000 membership fee and agreeing to purchase Subscription CTCs of 5 percent of the principal amount of each long-term loan, in addition to normal Loan CTC purchases, until the equity requirement was met.

The results of a joint CFC-NRECA member survey said a majority did not favor the change. "The recommendation was not intended to reduce the capital contribution of nonmembers but to collect those amounts in a manner than makes membership in CFC a more acceptable economic decision," Gill responded. "CFC now requires a substantial cash payment when a system applies for membership. The Member Advisory Committee's recommendation was to spread the cash payment over several years. The nonmembers would still be obligated to make the same investments and under the very same formulas that existing members were required to make when they joined CFC. We believe it is important for the program to have as many rural electric systems as possible in the CFC family."

The additional change was successful. By the end of fiscal year 1994, CFC had 1,039 members, including 833 distribution systems, 66 G&Ts, 71 service organizations and 69 associate members that joined under a program established in 1983. Less than 70 eligible systems had not yet joined CFC.

Member Advisory Committee Open Forums

After the district meetings concluded, the Member Advisory Committee held a series of open-forum hearings in four locations to receive additional feedback on the operating and policy issues under discussion. About 300 member system representatives attended. The Member Advisory Committee finalized its report on December 4 and presented it to the CFC board on December 9, 1991.

"The committee's final report is the product of one full year's effort," Chairman McGinnis told the board. "We firmly believe the recommendations, if adopted, will provide rural electrification with a stronger financing program and will prepare CFC to meet the challenges of the new decade and thus benefit our members and the communities in which they live."

The Member Advisory Committee report included the analysis of three investment bankers, Robert Edmiston of Lehman Brothers, Jerry F. Hilligoss of Merrill Lynch & Company and J. Scott Magrane of Goldman, Sachs. The committee had asked them to present their views on capital market access for rural electric cooperatives in the event that the REA loan and guarantee programs were sharply diminished or terminated. "CFC is eminently capable of supplying distribution member capital needs on attractive terms, even in the total absence of an REA loan or guarantee program for distribution cooperatives," they said. "It can do so without increasing substantially the interest expense burden of distribution cooperatives as a percentage of gross revenue."

The investment bankers' report also said that cooperatives should not fear the phaseout of the G&T guarantee program.

"Access to the capital markets on reasonable terms and conditions should be achievable for all but the weaker G&T systems," they said. "It is not a question of availability but cost. Cost should be defined as not only the incremental interest cost but also the financial standards necessary to attract capital from non-REA sources."

The bankers said that CFC might or might not be able to meet the future requirements of the G&Ts, depending on the quality and size of CFC's asset base and the amount required by any one borrower. "It is important for the G&Ts to be creditworthy and capable of attracting capital not only from CFC but the capital markets as well," they said. "It is in the self-interest of all G&Ts and their distribution members that CFC be as strong a finance company as possible so that its members can reap the benefits of economies of scale, size of asset base and risk diversification."

The report identified a significant impediment to market access. "Unless the REA permits *pari passu,* or absolutely equivalent, security interest to other investors, capital market access by the G&Ts will be constrained," they said. "It is inefficient to finance on an ad hoc project-by-project basis or to expect future providers of capital to provide funding to a G&T that would be subordinate to the security interest of the REA. It is standard practice for investor-owned and municipal-owned utilities to have lending agreements that have simple tests to be met as a condition of issuing additional *pari passu* debt."

The report noted that REA might not have the authority to provide equivalent security interests without additional legislation. The compensating benefit issue, cross-default issues and the sharing of remedies along with the related issue of personal liability under Title 31 of the United States Code, Section 3713 (the issue that precipitated the Wabash bankruptcy), could limit REA's ability to act. "In any event, the question of REA lien accommodation, in our view, is the single most important future financing issue facing the program," the bankers said. "It is a complex problem requiring careful professional analysis to determine an appropriate and equitable resolution."

Membership Approves Recommended Changes

The board unanimously endorsed the committee's report and recommendations, which were distributed to the member systems for action at the January 1992 annual meeting. By the time of the annual meeting, the members were well-informed, and the issues had been thoroughly discussed. The final action was anticlimactic. CFC's membership approved the policy changes and the four proposed bylaws changes.

Although several regions had passed resolutions favoring the Member Advisory Committee plan for transferring the revolving fund to a cooperative-controlled organization, the majority favored continuing the relationship with REA with a variety of suggested improvements. "The concept wasn't the pool of capital, because the difference in 6 percent money and 8 percent money is not going to make or break a distribution system," Gill said. "The concept was to have a source of commonality where all of the systems, big and small, urban, suburban and rural, would be getting a piece of the action." The concept was to create a force for program unity.

CFC's membership approved four key bylaws amendments at the 1992 annual meeting.

- Eliminate age requirement as director qualification
- Elect directors at district meetings; reduce minimum number of candidates to two; reduce percentage required to petition for slot on ballot
- Permit board to determine method of retiring patronage capital
- Eliminate requirement to issue patronage capital certificates annually

"It was misunderstood as a power play. It was misunderstood in the sense that a lot of our program's life and livelihood has been based on fighting political battles," Gill said. "The political process won out over the economic process."

In December 1991, the NRECA board approved a plan for improving REA that was adopted by the membership at the 1992 annual meeting. NRECA appointed three member committees to work with the REA staff to address the problems of overregulation, the REA mortgage and technical areas of engineering and construction problems.

Gill praised the work of CFC's Member Advisory Committee and NRECA's Cooperative Finance Study Committee and expressed support for the NRECA committee's recommendations on changes that would reduce the REA loan backlog and promote a better relationship with the agency. "At the same time, we should work to assure access to outside capital through more expedient and liberal lien accommodation policies on the part of REA," he said. "We should provide those systems desiring to end their relationship with REA with a means of achieving that objective. CFC is the single most significant financial commitment made by the rural electrification program. We at CFC do not take that commitment lightly. We are working with those systems that hope to improve the REA funding program, and we will assist those systems that are determined to extract themselves from the REA program."

Changes at REA

In 1992, CFC Senior Vice President and Loan Officer Richard B. Bulman provided testimony supporting the efforts of member systems to buy out their REA notes.

The efforts to improve the services and loan options offered by REA culminated in several pieces of legislation offering new opportunities for REA borrowers. The Rural Electrification Administration Improvement Act of 1992, which became law in October of that year, included a provision that allowed borrowers to buy out notes under their REA direct loans at a discount reflecting the government's cost of capital. However, a borrower choosing to buy its notes would be ineligible for additional REA loans for a period of 10 years.

In anticipation of the legislation, CFC developed software to assist members in evaluating the buyout of their notes. The interest among some systems was high, and CFC provided almost 300 analyses to members between March and August of 1992. REA did not publish its proposed regulations governing the buyouts until September 1993, and the final rules were not issued until March 1994. During this period, CFC revised its note buyout software several times to reflect anticipated changes in the REA rules and performed almost 500 more studies for member systems. It also made the software available to member systems who preferred to conduct their own analyses.

"If the rules had been issued earlier when interest rates were lower, the government could have saved money because the discounts would have been lower, and it could have saved money for the systems, who would have gotten lower interest rates," Richard Bulman said. "The same is true of the FFB refinancings. If that had been done faster, the systems could have gotten out of their high interest rate loans sooner. It was a lost opportunity to make borrowers more economically feasible."

By November 1994, 15 rural electric systems had bought out REA debt under the new procedures. Nine of the systems borrowed funds from CFC for this purpose, three used general funds and the remainder selected another lender.

The Omnibus Budget Reconciliation Act of 1993, which became law in August 1993, authorized all cooperative borrowers to refinance or prepay FFB loans with payment of a penalty. The pre-1984 loan agreement between cooperative borrowers and FFB allowed the borrower to prepay an FFB loan at least 12 years old by paying a penalty of one year's interest on the loan balance. After passage of the bill, FFB permitted refinancing within FFB on the same basis as prepayment. Loans less than 12 years old could be refinanced by paying one year's interest plus a fee based on the age of the loan. By January 1994, 19 G&T systems had repriced more than $660 million in 12-year-old FFB notes, for a substantial annual savings.

In 1992, CFC was second only to REA in terms of total long-term loan funds it had provided to distribution systems.

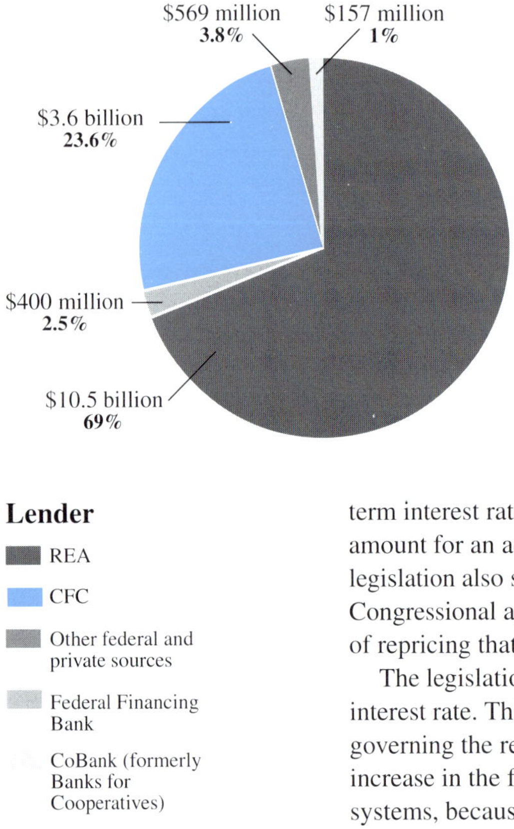

The new legislation authorized cooperatives to refinance loans under the post-1984 loan agreement by paying a penalty amount based on the difference between the loan balance and the market value of the loan and to select long-term or short-term maturities to take advantage of low short-term interest rates. The penalty could be included in the loan amount for an additional fee of 2.5 percent of the penalty. The legislation also said that the FFB notes could be repriced without Congressional appropriations, removing any ceiling on the level of repricing that REA could approve.

The legislation included a cap of 7 percent on the refinancing interest rate. This caused a delay in implementation of the rules governing the refinancing because of concern that rates could increase in the future. The delay was very costly to the G&T systems, because of increases in Treasury rates during the interim period. By fall 1994, few systems had repriced under this option.

Although the earlier FFB prepayment options required the borrower to obtain funds from another lender such as CFC, the options under the pre-1984 notes and the new legislation allowed the borrower to refinance through FFB. Thus, CFC's primary role was to advise the systems and to provide financing for the penalty. CFC staff helped more than 40 member systems evaluate the impact of the refinancing on total financing costs and evaluate options for paying the prepayment penalties. CFC also offered an unsecured intermediate-term loan program to finance the penalty in lieu of the longer term FFB penalty financing.

REA Moves to Market Rates

President Clinton signed the Rural Electric Loan Restructuring Act on October 1, 1993. This piece of legislation brought about the most significant changes in REA loans since the revolving fund was established in 1973. The new legislation retained the 5 percent interest rate for the neediest borrowers, defined as those with a significant rate disparity compared to neighboring utilities, customer income lower than the state's average or extremely high rates. For other systems, the interest rate was tied to the current market yield on municipal bonds of equal maturity. The municipal rate was selected to equalize government assistance to tax-exempt municipal utilities and cooperative utilities. Systems would be allowed to choose a fixed interest rate for maturities up to 35 years; REA established a minimum maturity of three months. The interest rate would be capped at 7 percent for systems with less than 5.5 consumers per mile of line or rates exceeding the state utility average and a consumer base whose per capita income was below the state average.

The Restructuring Act initially provided higher loan and guarantee levels, which were expected to reduce the substantial backlog in loan applications. It also reduced or eliminated required operating reviews and approvals under some circumstances. In addition, it allowed REA to provide loans and grants for some rural economic development activities.

A further organizational change occurred in October 1994 when various federal programs, including those administered by REA, were consolidated under the Rural Utilities Service, a new branch of the Department of Agriculture.

The revisions in the REA program provided new opportunities, but the changes also presented new challenges. Historically, most distribution systems had borrowed more than 70 percent of their debt on a fixed-rate basis from REA. Only the portion from the supplemental lender was subject to some degree of variability.

"Many of our members have become used to a certain type of operation," Gill said. "None of them have ever operated in the market before, and REA has thrust market decisions on them with the variable rate. We are now trying to educate the members about the various options." A potential problem was that some members expected CFC to insulate them from market changes, something CFC could not do. "When you are working for one *member* you may be working against the *members.* They may make some erroneous decisions from time to time, and they are going to have to learn to live with those decisions. The saving grace is that interest at the distribution level will not make or break the cooperative."

Thus CFC accepted a new job: helping members develop an interest rate strategy.

"CFC has provided members with information regarding interest rate history, trends and key financial indices that may impact interest rate markets," Sheldon Petersen said. CFC expanded the interest rate information and analyses presented in its publications and set up an interest rate telephone line. A new electronic bulletin board system also provided interest rate data and allowed members to download rate information as often as they wanted.

The member systems responded positively to the opportunities offered. "The future is going to be CFC as a partner with the cooperatives, not a banker to the cooperatives," Gill said.

CFC and NRECA Enhance Relationship

The long months of discussion over future financing prompted CFC and NRECA to explore ways in which they could work more closely together for the good of the program. One of the options considered by the Long-Range Study Committee was to make CFC a department within NRECA. In the end, the arguments in favor of a separate organization were more persuasive, and CFC, with great support from NRECA, was established as a separate organization. Over the years, there were differences in viewpoint between the two organizations.

"The difference between the two organizations is the political purpose versus the economic purpose," Gill said. "For political purposes we do certain things that are absolutely contrary to economic purposes." For example, NRECA might highlight the weaknesses in the program to gain support for continuing subsidized interest rates on REA loans, whereas CFC would highlight the strengths in marketing rural electric credit on Wall Street. "It was something both organizations had to deal with. We did not have a conflict of principle or of supporting cooperatives, but sometimes we did have a conflict of approach," said Gill.

***In May 1994, the CFC board declared a one-time rebate of $8 million to performing borrowers who paid interest on loans during the first eleven months of fiscal year 1994. CFC Governor Charles B. Gill* (left) *and President Bill McGinnis* (center left) *presented a check for NRECA's share of the rebate to NRECA President J. C. Roberts* (center right) *and Executive Vice President and General Manager Glenn English* (right).**

Once other lenders began to target the rural electric program, NRECA took the position that, because not all of its members were CFC members and some of its members preferred other supplemental lenders, it should remain neutral. It included CoBank and investment banks in its programs, sometimes to the exclusion of CFC.

"Over time, NRECA felt it was their function to get as much financing from everybody as they could," Gill said. CFC's view was that NRECA and CFC were essentially owned by the same people. Although the members were certainly free to choose any supplemental lender, CFC believed it was inappropriate for NRECA to actively promote its competitors.

"NRECA was our mother. They started us. At the same time, our mother was sponsoring forums that included Smith Barney and CoBank discussing financial options," Petersen said.

CFC became increasingly concerned about legislative issues that affected its ability to provide financing to its members. From the beginning, CFC agreed that it would not initiate lobbying activities. However, there were times when NRECA was unable to adequately represent CFC's interests because of its policy of neutrality toward supplemental lenders.

CFC and NRECA addressed the problem in 1989 when NRECA added an additional legislative specialist to its staff to track banking and financial issues. CFC agreed to fund the position. That helped, but CFC still had concerns.

Finally, Governor Gill and President Bill McGinnis raised the issue directly with NRECA. "Either CFC is your supplemental lender, or it isn't. Either you support us, or you don't. If you don't, we are going to have to do some things on our own," they said.

"It was absolutely the right thing to do because it broke the ice for all of us," Petersen said. "When they were confronted and asked, 'Are you with us or not?' their response was, 'We're with you.'"

The two organizations began to make some basic changes in the way they worked together. NRECA General Manager Bob Bergland and NRECA President J. C. Roberts, general manager of South Plains Electric Cooperative in Lubbock, Texas, provided leadership in recognizing what needed to be done. As NRECA president, Roberts was adamant that the rural electric program as a whole would be better served if NRECA and CFC cooperated in their activities.

CFC Designated Primary Supplemental Lender

The NRECA membership passed a resolution at the 1994 annual meeting designating CFC as the primary supplemental lender. The two organizations found many ways to cooperate in providing services to their joint membership, such as holding joint seminars on the REA note buyout program, sponsoring a task force on rural water infrastructure together and jointly operating an electronic bulletin board.

"We are both coming to recognize that we have more to gain by acting collectively than we do by acting separately," Gill said.

Matching Debt and Loan Portfolios

While the program at large was contemplating REA's role in future financing, CFC continued to refine its own financing options. One of the issues was the need to increase variable-rate funding to match a shift by the membership to more variable-rate loans. CFC offered members a great deal of flexibility in converting fixed-rate loans to variable rates, and as interest rates dropped in the early 1990s, many took advantage of the opportunity to convert loans with relatively high fixed rates to a lower cost option.

"We try to match our loan portfolios as closely as we can, matching fixed-rate debt with fixed-rate loans and variable with variable," said Steven Lilly, CFC's chief financial officer. In the early 1990s, more that $1.4 billion in loans was converted from a fixed interest rate to a variable interest rate, whereas about $400 million was converted from a variable rate to a fixed rate. "There was a net change of about $1 billion, plus almost all the new advances were on the variable-rate side. We needed to fund that loan growth and conversion with variable-rate securities to minimize our interest rate risks."

CFC's commercial paper program was a primary source of variable-rate funds. It was backed by bank lines of credit and revolving-credit agreements, long-term commitments to lend short-term funds. CFC rarely borrowed from those credit sources, but they were essential to CFC's strong commercial paper credit rating. CFC was also required to maintain backup lines of credit for the variable-rate tax-exempt bonds issued under its guarantee.

Although the 1990 downgrade had very little impact on CFC's ability to raise funds or on its cost of money, the overall credit standing of the rural electric program, along with a domestic banking crisis, affected CFC's ability to increase the amount of credit available from domestic banks. Some banks simply did not want to have any credit exposure to the rural electric program at that time, and the banking industry itself was experiencing hard times because of failed real estate investments. In addition, federal regulatory authorities put limits on the amount of exposure the banks could have to one entity.

"It was obvious that we were not going to be able to get the same amount of credit," said Truman Brandt, who was CFC's chief financial officer at the time. "First the banking industry went down. Then our problem loans started to come in, when Moody's downgraded the entire industry. Several large banks determined that they didn't want to have anything to do with electric cooperatives, particularly those with direct credit exposure to troubled borrowers. Two or three others said they did not want to make any credit available to finance companies due to competitive factors."

CFC's finance group implemented a two-pronged strategy to be sure adequate credit was available for the member systems. The first prong was an aggressive marketing campaign. "We met on a weekly basis with potential investors, actual investors and our banks and very frequently with the rating agencies to keep them abreast of the changes, which were all very positive," Lilly said. "We relayed to each one of these firms our degree of comfort with our ability to absorb the risk associated with Colorado-Ute and Deseret. We laid out for them the worst case scenarios and showed how we could meet them. We felt all along, and we relayed this to the investors, that we would not have a loss in Colorado-Ute, but we showed the capital markets that even if Colorado-Ute didn't pay cash on an ongoing basis, we would have sufficient financial flexibility to cover the shortfall in the net funding cost. Our commitments to Wabash and Deseret were primarily guarantees and periodic payment of principal and interest, so we had a reserve built up and sufficient time to continue adding to the reserve to cover any worst case situations."

One of the financial analysts' biggest concerns was that the rural electric situation was similar to that of the real estate industry, where assets in foreclosure had no source of revenues. "We always had an expectation that there would be a revenue stream from the borrower. Although it might not be 100 percent of the requirement, it would be substantially more than zero," Lilly said.

Developing International Credit Sources

The second part of CFC's strategy was to develop new sources of bank credit. "Because the domestic banks were getting hit hard, we had to put major emphasis on developing foreign banks, particularly the European banks, because we had a problem with some of the Japanese banks because of the Sunflower situation," Brandt said. "The European banks had more credit available, didn't have to make as many write-offs or build up their loan reserves and were in much better shape with their credit ratings." Although the overall average credit rating of the major domestic banks had fallen to a low A in 1990, the foreign banks CFC was dealing with maintained triple-A or strong double-A ratings. Many of the U.S. banks were trying to reduce overall expenses to improve profitability and conserve capital. Some were abandoning national accounts, whereas others were reducing their credit commitments. "There were a lot of other companies that we were competing with to obtain capital from our domestic banks, and the troubled U.S. bank situation also intensified the competition for available credit from foreign banks, particularly the European banks," Brandt said.

The European style of banking differs from that in the United States in that it relies much more on personal relationships. "Part of the reason for that is that they don't have the same information systems available to the public that we do," Brandt said. "They don't have a Securities and Exchange Commission, for example, so there is not a lot of data available on many of the companies. We try to reinforce a personal relationship with them by sharing worldwide outlooks on financial policies and philosophies to assure them of the quality of our organization's management."

CFC called regularly on the European banks. A typical trip would involve visits to more than a dozen banks in Great Britain, the Netherlands, Germany, Austria, Switzerland, Italy and France. Their reasons for interest in CFC varied. Some wanted to develop positions in the United States. Banks that already had positions in the United States were interested in the different types of financings CFC did and its potential requirements. "They were interested because we had longer term plans that we might sell securities in Europe as well. They could possibly be an underwriter in their own country on a special basis," Brandt said.

The effort paid off. By 1990, foreign banks were providing almost $1 billion of credit to CFC through bank lines and revolving-credit agreements. In 1993, CFC negotiated two new revolving-credit agreements, including a $2.03 billion three-year facility and an $870 million annually renewable facility, that attracted 52 banks, including 31 international banks. The international banks provided well over half of the total credit. CFC's success in negotiating this type of credit facility demonstrated the wisdom of its initial decision to establish its relationships with the banks directly, rather than working through a lead bank.

CFC further expanded its capability to raise short-term funds when it established a bid note program, approved by the board in April 1992. A bid note is created when a bank issues a loan to a company and then sells some or all of its interest in the loan to other investors. The maturities on the notes are similar to that of commercial paper, but bid notes do not require back-up lines of credit. "The basic changes in the capital markets were affecting the availability and cost of the backup we needed for commercial paper," Brandt said. "We needed to develop alternative sources for variable-rate funding." The new program

In June 1991, CFC entered into a new $1.75 billion, three-year revolving-credit agreement with Morgan Guaranty Trust Company of New York as agent and 37 other banks. The credit facility provided credit support for CFC's commercial paper program and guaranteed remarketable pollution control bonds. Pictured,* left to right: *CFC Director of Banking Relations Kimberly P. Armstrong, CFC Chief Financial Officer Truman L. Brandt and Morgan Guaranty Trust Vice President Thomas B. Nichols.

gave CFC a competitive, and potentially lower, all-in cost of capital and an opportunity to further develop relationships with its banks.

Restructuring CFC's Long-Term Debt

At the same time it was expanding its short-term credit facilities, CFC was also restructuring its long-term debt. "We began pre-paying our collateral trust bonds in the late 1980s because it was favorable to do so," Brandt said. CFC was able to repurchase all of the bonds issued in the early 1980s at interest rates higher than 10 percent as well as Series F, which carried an interest rate of 9.75 percent, Series Q at 9.375 percent, Series R at 9.625 percent and Series S at 9.85 percent.

Favorable market conditions allowed CFC to issue new long-term debt at attractive rates and terms. Series T, a 1990 $150-million seven-year issue with a coupon rate of 9.5 percent, was followed in 1991 by Series U, a $150-million seven-year issue priced at 8.5 percent and Series V, a $150-million 30-year issue

CFC collateral trust bonds have a strong credit rating history. Series A through V were issued under the 1972 Collateral Trust Bond Indenture.

Series	Issue Date	Due Date	Amount (millions)	Interest Rate (%)	Credit Rating Moody's	S&P	Fitch
A	December 1972	2007	$ 50	7.4	A	A	NR
B	October 1974	1981	100	10.5	A	A+	NR
C	January 1975	1985	100	8.95	A	A+	NR
D	October 1975	1985	100	9.125	A	A+	AA
F	March 1979	2009	100	9.75	A	AA	AA
G	August 1979	2009	100	9.75	Aa	AA	AA
H	September 1980	1990	50	12.625	Aa	AA	AA
I	September 1980	2010	50	13.5	Aa	AA	AA
J	January 1981	2011	100	14.75	Aa	AA	AA
K	May 1981	1991	100	14.875	Aa	AA	AA
L	August 1981	1991	100	15.75	Aa	AA	AA
M	November 1981	1988	100	14.625	Aa	AA	AA
N	July 1985	1995	100	10.5	Aa2	AA	AA*
O	March 1986	2016	100	9.0	Aa2	AA	AA*
P	October 1987	1992	100	9.25	Aa2	AA	AA*
Q	June 1988	1995	150	9.375	Aa2	AA	AA*
R	January 1988	1996	100	9.625	Aa2	AA	AA*
S	April 1988	1996	100	9.85	Aa2	AA	AA*
T	May 1990	1997	150	9.5	Aa3	A+	AA*
U	February 1991	1998	150	8.5	Aa3	A+	AA-
V	September 1991	2021	150	9.0	Aa3	A+	AA-
1994-A+	September 1994		150	Variable	Aa3	AA	AA
Downgrades:							
	April 1990				Aa3		
	May 1990					A+	
	July 1991						AA-
Upgrades:							
	October 1975					AA	
	August 1979				Aa		
	May 1992					AA-	
	May 1994					AA	AA

Moody's, Moody's Investors Service; S&P, Standard & Poor's Corporation; Fitch, Fitch Investors Service.
*Fitch did not rate this series on the original issue date.
+Issued under the 1994 Collateral Trust Bond Indenture.

priced at 9 percent.

The favorable changes in the cost of long-term debt allowed CFC to reduce the interest rates on its loans correspondingly. The standard long-term fixed rate fell from 10.5 percent at the beginning of 1986 to 9.5 percent at the beginning of 1989 to 8 percent in the middle of 1992.

CFC devoted a substantial portion of its educational efforts to the rating agencies. Although at that point CFC had not adopted a formal policy on credit ratings, it had long had an informal goal of maintaining double-A ratings across the board. The 1990 downgrade did not hurt access to and cost of funds: it hurt CFC's pride.

The CFC rating was affected by the rating agencies' overall perception of the rural electric program. "Rating agency analysts base approximately 60 percent of their evaluation on industry-related issues and 40 percent on analytical evaluations of achieved results," Brandt said. "They were forecasting a less rosy future for the rural electric industry and that the role of REA would decline. They were concerned about what the credit quality of the industry would be if REA weren't there, because they felt that REA was the only regulatory influence."

Other concerns included the distribution members' willingness to accept responsibilities under their wholesale power contracts, the adequacy of rate increases and member guarantees of G&T financings. The potential of an REA note buyout program and its impact on CFC was also an issue. "These unknown factors helped determine CFC's risk profile for investors," Brandt said.

CFC continued to emphasize the high asset quality of its diverse loan programs, collateral security, liquidity backup, loan loss reserve fund, financial ratios and overall membership equity investments and industry support. "Beyond that, we emphasized the ability of CFC and the industry to self-regulate," Brandt said.

Most financial institutions, including banks, savings and loan companies and credit unions, are regulated to some degree. Many finance companies, including CFC, are not, partly because they do not have insured depositors. "We believe self-regulation is better for a company like CFC than any regulatory body could ever be, because when we take a loss, our members lose something," Lilly said. "That is more important than having a regulatory body say you should have a certain amount in the loan loss reserve and consider putting this loan in the nonperforming loan pool. Oversight by the members who have a vested interest in the bottom line is stronger than any external regulation could be."

CFC established limits on credit and eliminated programs, such as the unsecured front-end financing program for G&Ts, which had proved risky in the past. Furthermore, it emphasized to member systems their role in maintaining CFC's creditworthiness. "CFC's credibility is influenced greatly by the reputation of its members," Brandt said. "Self-regulation that includes quality management and training will allow cooperatives to maintain a long-term relationship with investors based on honor and trust. Managers and boards must understand fiscal responsibility to all affected parties and should be encouraged not to take a short-sighted approach to directing their cooperative."

Efforts Bring Ratings Upgrade

CFC's efforts came to fruition on May 28, 1992, when Standard & Poor's upgraded its credit rating on CFC's collateral trust bonds to AA- from A+ and the rating on commercial paper to A-1+ from A-1. Standard & Poor's attributed the upgrade to CFC's enhanced operating posture and financial standing resulting from the completion of the Colorado-Ute bankruptcy reorganization, as well as a general improvement in CFC's asset quality.

At about the same time, Moody's and Fitch Investors Service reaffirmed their ratings of Aa3 and AA-, respectively, on CFC's collateral trust bonds and PRIME-1 and F-1+, respectively, on commercial paper. CFC had regained its objective of being rated double-A by all three rating agencies and maintaining the highest rating on commercial paper.

NCSC continued to service the transactions that were executed in the 1980s and to provide useful services in unique situations. Unlike CFC, NCSC was able to lend to for-profit companies, so it provided financing for some for-profit affiliates of CFC members. With no TIER requirement, it could provide loans at a lower rate than CFC in some situations. For example, NCSC furnished debt to Southern Maryland Electric Cooperative to finance a combustion turbine project. A unique aspect of the financing was that it included a swap based on NCSC commercial paper. "NCSC is still serving a purpose," Petersen said. "As the picture changes as to what types of financing are available to cooperatives, it will develop new programs in the future."

A New Indenture

With the issuance of Series V, CFC's long-term funds were sufficient to meet its long-term funding needs for an extended period of time, so the organization took the opportunity to develop a new indenture for long-term issues.

The original indenture was structured when both CFC and the cooperative utility business were new to the capital markets. Because of that, it included comforts to investors that were no longer needed and no longer provided a benefit to CFC. "The indenture served us well for a long time, but it also cost us money," Lilly said. "We have proven ourselves over the last 25 years. We don't need the belt and suspenders that were in there to protect investors."

The new indenture eliminated a costly debt service reserve fund and offered collateral from distribution systems only. "We think that the bulk of our business is going to be in distribution system financing," Lilly said. "We have other funding products to raise capital for our other core lines of business. The investors have an overall claim on CFC, so we can still get a high rating and good reception in the market. If we need to fund the G&Ts, for example, we could do that with unsecured medium-term notes without the pledge of specific assets. We hope that the new indenture, along with the improvements in the credit quality of the portfolio, will elevate us to another notch on the rating scale to a mid-double A." The new indenture was also expected to reduce costs by $300,000 to $400,000 each year.

The credit upgrade came in May 1994 when both Standard & Poor's and Fitch raised CFC's rating on collateral trust bonds to AA. "CFC's overall general asset quality and loan concentration continues to improve," Standard & Poor's said. "Also incorporated in the rating is CFC's expected ability to retain its core base of borrowers."

The rating agency had some comments on competition in the industry. "Standard & Poor's believes that the cooperative industry will feel the effects of increased competition in a variety of ways, including consolidation, the need for timely access to and availability of low-cost capital, and pressures to lower rates," it said.

The upgrades placed CFC's ratings at the upper level for domestic finance and electric utility companies. "The upgrade is concrete proof of our members' improved financial strength," Lilly said.

The first issue under the new indenture, Series 1994A, took place on September 16, 1994, when CFC issued $150 million in collateral trust bonds at a floating rate based on monthly London Inter Bank Offered Rate indexes. The issue provided another secure source of funding for variable-rate loans. An interest rate swap tied the cost of the issue to commercial paper rates.

CFC's financial strategy for the 1990s was based on the interaction of diversified funding sources and a strong credit rating. "A strong credit rating enables us to have very diversified funding sources," Lilly said. "Diversified sources of funding enable us to lower our costs, because we have more investors and banks competing to provide us with financial services and products and credit. All of that is then passed through to our members. CFC's diverse financial resources minimize the requirement for borrowers to have multiple lenders."

CFC also continued to emphasize the long-term nature of its relationships with institutions and investors. "We are a very high-quality credit, so we can negotiate very low fees. Many institutions don't see us as an opportunity to make a significant return, but it is a steady return," Lilly said. "What we sell is that CFC is going to be there for 10 or 20 years, or longer. Ultimately, banks need a core level of business to sustain them through good and bad times. We try to align ourselves with institutions from that relationship-core business standpoint."

Although the dominant share of CFC's total funds was raised in the United States, it continued to nurture its international relationships. "We have looked very closely at all the European products, including Eurocommercial paper, Euromedium-term notes and Eurobonds," Lilly said. "We have also been introducing CFC to the major life insurance companies and investment funds in Japan for the potential placement of fixed- and variable-rate intermediate-term securities."

With strong credit ratings, a new indenture and diversified funding sources, CFC had positioned itself well to provide for its members' needs throughout the decade.

A New Vision for CFC

The presentation of the Member Advisory Committee report and the approval of the bylaws changes was not the end of CFC's strategic planning efforts. In many respects it was the beginning. "We were back to where everybody understood where REA was going to go again, and finance was no longer a hot topic," Petersen said.

Attention returned to the changes in the electric industry. "CFC has for the last several years been concerning itself with what we need to do to be ready for tomorrow," former Administrative Officer Nicholas Walsh said. "One of the things that we hear is that the electric utility industry needs to find a way to be more competitive. If being economically viable and competitive is a concern of our members, it should be a concern of ours. The co-ops are changing, we are a part of it, and we need to change."

With the help of an outside facilitator, the staff and board held a series of meetings to evaluate the way CFC operated. The group looked at the programs and services CFC provided to members and the methods it used in providing them. "We found a lot of things that concerned us," Walsh said. "We knew we could do better."

CFC had planning sessions in the past, but they did not precipitate the degree of introspection and consideration that this series of meetings did. "It was not until we started to be serious about it, bringing in an independent person to help the staff and the board, did we realize what we were about and what was going on," Richard Bulman said. "Those sessions where we were able to be frank and candid helped change the company. There was a sense at the end of it that everybody had seen the same vision."

The vision was formalized in a strategic plan adopted by the board in May 1993. The plan was not created overnight, and the changes recommended took place over time during and after the period in which the plan was developed. CFC defined its mission and identified a strategy for fulfilling the mission. The plan set priorities and outlined specific actions CFC could take to meet its goals.

CFC's purpose had not changed since its incorporation almost 25 years earlier. It was still to provide an assured source of low-cost capital and state-of-the-art financial services for its member systems. CFC rededicated itself to being the primary supplemental lender to the rural electric program while offering opportunities for systems desiring to terminate their relationship with REA. However, in meeting individual systems' needs, CFC also needed to work to strengthen program unity.

"Unity without uniformity. Be competitive. Be an agent of

change. It is a movement, and the 1990s have brought CFC to that movement," Gill said. "The strategic plan was very helpful in bringing the board up to speed and having the board bring the staff up to speed on what some of the issues really were."

REA would remain the primary source of long-term financing for most of the distribution systems, but the note buyout program offered an opportunity for systems that wished to do so to end their relationship with REA. This would benefit the remaining systems by reducing the demand for limited REA funds.

The strategic plan identified the note buyout business as an important opportunity for growth in CFC's core business, but the board and management recognized that member systems now had many options. The services CFC offered had to be competitive with alternatives if CFC was going to continue to meet its members' needs and attract the buyout business. "I had been preaching for a long time that the members of a cooperative will not be loyal simply because you performed a mission 50 years ago," Gill said. "We applied that to CFC and asked, 'Are we providing a product in an acceptable fashion at a price people are willing to pay, or are we relying on old loyalties?'"

More Flexible Loan Programs

With the new indenture well underway, CFC was positioning itself to continue to raise funds at a competitive cost. Changes in the loan programs were intended to ensure that CFC offered the best loan products. In September 1992, the board approved a new method of setting interest rate maturities, effective in October. In addition to the variable interest rate and the standard seven-year fixed rate, the program was expanded to include mid-term fixed interest rates for periods of 1 to 6 years and extended-term fixed interest rates for periods of 8 to 30 years. The new options gave member systems great flexibility in selecting pricing mechanisms to fit individual needs.

In the closing months of 1993, CFC implemented several more changes. It reduced or eliminated the up-front equity requirement for Loan CTCs on secured long-term loans for many borrowers. The maximum CTC requirement for distribution systems was reduced from 7 to 3 percent, and many members were no longer required to make any equity contribution as a condition of loans. This directly reduced the cost of the loan to the member. In keeping with the recommendations of the Member Advisory

Committee, the change would, over time, also result in an increase in CFC's leverage ratio.

CFC restructured the way it retired patronage capital by lengthening the rotation cycle to up to 15 years and adopting the percentage method of retiring patronage capital. CFC patrons would receive half of the patronage capital earned in a given year immediately with the remainder refunded over a 15-year period. "If we can give patronage capital back currently through the percentage method, it effectively reduces the cost of the next loan, as opposed to waiting six or seven years to reduce the cost on a future loan," Gill said. It provided an immediate reward to members who supported CFC. For those members who had tended to discount patronage capital in their analysis of loan alternatives, it also demonstrated CFC's commitment to return it.

A new program of interest rate discounts was implemented for qualifying borrowers for long-term fixed-rate loan volume above a specific threshold. "Differential pricing on loan volume can be justified on the basis that we can service one $80-million loan at considerably less cost than we can service 80 $1-million loans," Gill said. "The one-eighth point that we give them will come back to us in the sense that it will drive down the administrative costs by that much. It won't cost the other borrowers anything." And the other borrowers would benefit from the growth in CFC's asset base.

CFC also implemented rate discounts on newly priced fixed-rate loans based on financial performance. At the time the decision was made, there was a difference of about 10 basis points in the cost of long-term funds to double-A–and single-A–rated companies. Better financial performance by members allowed CFC to maintain a higher rating. "The discount for performance showed that CFC recognized the system's value to the loan portfolio," Bulman said.

Finally, CFC began to establish interest rates on a weekly basis. It gave members more flexibility in structuring their individual loan portfolios to match their needs. It also allowed CFC to be more responsive to changes in the financial markets.

"There was something in this program for just about everyone," Gill said. "Our intent was to provide borrowers with an attractive package that would demonstrate CFC's commitment to our members where it counted most—on the bottom line."

Streamlining Loan Approvals for Fast Access to Funds

For many systems, service was an even more important consideration than cost. "Along with the more aggressive loan pricing, quality of service and responsiveness are very important," Hedberg said. "They are two of the ways our members are going to judge us."

The board increased the governor's authority to approve loans. For example, the board delegated the authority to approve all loans for REA note buyouts to the governor. "We needed CFC's management to have more flexibility of movement, so that the staff would have greater latitude to move and react to situations," Gill said. "We were going to lose business unless we convinced the membership that we were moving and we were competitive."

The need for quick access to funds was one of the issues driving the interest in the REA buyouts. Even with a reduced backlog, it was taking REA a year or more to process loans. The more rapidly growing systems were concerned about having enough loan funds to meet their member needs.

"We have a number of borrowers whose capital needs are so large that they need a loan of $20 or $30 million," Bulman said. Getting a loan of that size approved by REA can take months or even years. In the eternal debate over the importance of access to funds versus cost of funds, those systems wanted access to funds, if not from REA, then from other lenders. Only when funds were available could price become an issue. "Under the mortgage, REA decides whether a system can have a lien accommodation. As long as that is the case, some lenders are going to be unwilling to accept the uncertainty about whether they will get collateral, how long it will take and whether they will close or not—all the uncertainties that go with a government agency approval process. The driving force behind borrowers that want to pay off REA is that they want to have access to capital without REA approval," said Bulman.

Quick Action for Telephone Acquisitions

Rapid loan approvals were especially important to telephone borrowers competing for acquisitions. They needed to be able to offer bids in the time frame dictated by the selling company and to demonstrate their ability to pay. Through CFC, RTFC committed that it would process such loans within 30 days—more quickly if necessary—and provide whatever documentation was needed for the system to pursue the acquisition.

CFC restructured its loan approval process to provide a faster turnaround. It reduced the amount of information required for loan applications and annual reports. It substantially decreased the approval time for concurrent loans and prepared to process other loans even more quickly. "If you are not an REA borrower, it is a very simple process," Bulman said. "We do not have to wait for a lien accommodation. We can give a very fast turnaround."

CFC could also offer a wider range of products to members who were not REA borrowers. "With REA, we can offer rates for up to 35 years, but when we get past straight loans that amortize, it is more difficult to go through the approval process," Bulman said. Non-REA borrowers have more options in structuring a loan portfolio. They may select loans of different maturities and different funding mechanisms. They can take advantage of interest rate swaps and futures. They can select a series of non-amortizing loans with different maturities, which function similarly to bond sales. "If you are not an REA borrower, you can have almost unfettered access to funds as long as you meet certain predetermined financial ratios."

To become more competitive, CFC also had to become more efficient. It cut administrative costs from 37 to 25 basis points. The number of staff positions in the budget fell from 204 in 1991 to 168 in 1995. "We didn't fire anyone, but if a job came open, we asked ourselves it we really needed to fill it," Gill said. "We asked if we really needed to do the function that the job did or if we needed only certain elements of the job function that someone else could perform."

CFC's organizational structure changed over time to consolidate some departments. "The whole idea was to create a management team instead of having individual players," Gill said. "It's an attempt to become more functional, more reactive and to emphasize member services more. By having fewer senior staff people, we forced down the decision process."

Changing the Corporate Culture

Changing policies and procedures and corporate structure was not enough to change the way CFC interacted with its members. It had to change its people. "The strength of the 1990s is CFC's reactive mode to the membership, the movement away from the institutionalization of the financing of the corporation to a member-friendly institution," Gill said. "Many of the changes we have made, such as the change in the way we elect directors and the changes in the loan programs, were in response to members. We are being responsive in the competitive sense, being more member friendly, emphasizing marketing and emphasizing a different, nontechnical approach with the members. We are letting the members run their business. We're not REA anymore. We're their bank and their financial partner."

The CFC board of directors honored CFC's first governor by naming the new headquarters facility in Herndon, Virginia, the J. K. Smith Building.

CFC management recognized that there would have to be changes in CFC's corporate culture if the staff was to be successful in finding new ways of interacting with member systems. "The move to Herndon helped," Gill said. CFC sold its Georgetown property at a substantial gain and moved into a new headquarters facility in Herndon, Virginia, in the fast-growing Dulles access corridor in September 1990. Although other CFC moves demonstrated CFC's growth, this move demonstrated a change in philosophy. "There was a lot of turnover in people because of the move. That gave us an opportunity to infuse new philosophies. The transition allowed us to change how we operated internally as well. It was a whole different thought process out here, a positive new environment."

CFC designated member service as its highest goal and emphasized the responsibility of every employee to contribute to it. "CFC is a member service organization that happens to provide financial services," Petersen said.

CFC also acknowledged the need to be competitive. "When I first came to CFC, there was an attitude that CFC did not need to be competitive. There was no one to compete with," said CFC Director of Member Finance Martin Crowson. "Now management is much more member-service sensitive and open to new ideas. The staff is really sincere in trying to make these changes a success

2201 Cooperative Way

CFC purchased land for a new headquarters facility in the Woodland Park corporate office center in Herndon, Virginia, near Dulles International Airport in February 1986. The 12-acre site, which fronts the Dulles Airport Access Road, is about 16 miles from Washington, D. C., in one of the fastest growing corridors in the metropolitan area. At 230 acres, Woodland Park will provide more than six million feet of office space when it is fully developed.

Construction on the new buildings started in April 1989. CFC engaged Davis & Carter, P.C., as the base building architect, Leo A. Daly as the interior architect and Sigal Construction Corporation as the general contractor. The 141,503-square-foot facility occupies only four acres of the site, leaving room for future growth or development with other cooperative organizations. In addition to regular office space, the facility includes a board room with a state-of-the-art sound system, a fitness room, kitchen and 140-seat cafeteria, and a 4,728-foot multipurpose room seating 200, which can be used by member systems for meetings and seminars. A 350-space parking garage with 35 additional surface spaces provides ample parking for employees and guests. The attractive glass and polished granite building overlooks a 3.5-acre pond.

CFC initially occupied about half the space. The National Rural Telecommunications Cooperative and other tenants shared the remaining portions of the building.

Reduced operational costs and improved access to staff resources were the two major issues that influenced the decision to move. Parking, for example, had become very expensive in Georgetown, and support staff and young professionals were unwilling or unable to afford the high cost of commuting downtown.

"We did everything we could to make the new location easy for the members to visit," said former Administrative Officer Nicholas J. Walsh. CFC provides a van to transport visitors to town or the airport and has made special arrangements with a local cab company to provide immediate service.

Development of this modern facility was one of the steps CFC took to prepare to meet its members' needs into the next century.

in the sense of having a solid member constituency and finding ways to be helpful and cost effective."

CFC encouraged its employees to find new ways of working with the member systems and each other. There were training sessions to make them more aware of others' viewpoints. "Then we did some cross-departmental pollenization, putting task teams together across departmental lines to make an issue everybody's issue and not just one department's issue," Gill said.

Each year, some CFC employees spend time at a member cooperative to gain firsthand experience in the everyday routines of providing electricity.

There was a new program to send CFC employees to member systems for a week or two to learn about billing, work with the line crews, interact with the consumers and get firsthand experience in the everyday routines of providing electricity in a cooperative environment.

At one point, CFC even reviewed all of its standard letters. "If there was a word that was negative, we changed it," Gill said. "Instead of 'You are required to' we said 'We would like to receive' or 'If you could help us by.'"

CFC changed the way employee performance was evaluated to underscore the employees' role. It implemented a new performance appraisal program based on results, that is, what the employee was supposed to do, and behavior, how the employee interacted with others, including member systems.

Then management asked employees for suggestions about things that CFC should do differently to better fulfill its responsibilities. "We had a change in the direction of CFC as a result of the strategic planning sessions. We recognized that we can get more from our employees by empowering them, by letting them be problem solvers instead of solving the problem for them," Bulman said.

"CFC has changed internally," John List said. "We've tried to break down these artificial barriers that we had erected over time about there being a loan department and an administration department and a legal department and a finance department, who could touch this and who could talk to that person. We've tried to get rid of the internal turf defense mechanisms that build up in any corporate organization."

The membership responded positively to CFC's evolution. "The changes we have made have gone far beyond what anybody ever expected us to do in order to be competitive," Petersen said. "People are acknowledging that, and some of our biggest critics have become some of our biggest cheerleaders because of the changes we have made."

Gill Retirement Brings One More Change

There was one more change in store for CFC. On May 25, 1994, Charles B. Gill announced to the board of directors and staff that he would be retiring as governor of CFC at the February 1995 annual meeting. Under his leadership, CFC's assets grew from $1.3 billion in 1979 to almost $7 billion in 1994. The loan and guarantee programs expanded to provide rural electric systems with the most sophisticated financing tools available. The sources of financing evolved from member equity, collateral trust bonds and commercial paper to a diversified funding program that accesses international markets as easily as domestic ones. Gill guided the organization safely through the some of most difficult times in the history of the rural electric program. Most important of all, he left the organization well-prepared to face the future.

On December 5, 1994, the CFC board of directors announced that Sheldon C. Petersen had been selected to succeed Gill as

CFC's loan and guarantee programs have provided its member systems with billions of dollars of capital at favorable rates.

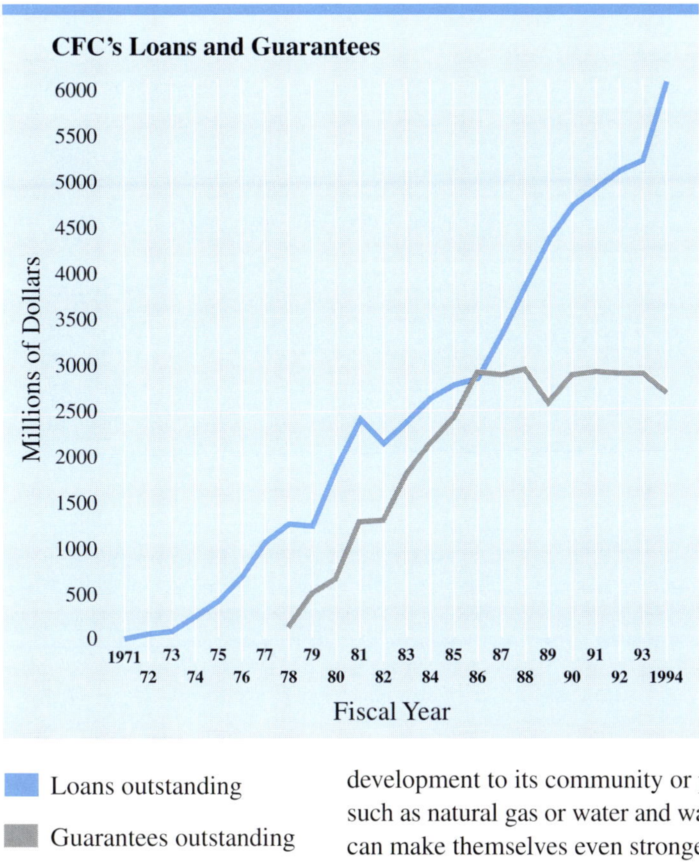

CFC's third governor and chief executive officer. "The board has the utmost confidence in Sheldon's ability to continue the record of member service and financial soundness that has been CFC's trademark for 25 years," said President Bill McGinnis.

The changes that came about through the strategic planning process positioned CFC to meet its members' needs beyond the year 2000. Helping them become more competitive will be a driving force in CFC's future programs and operations. With the implementation of the Energy Policy Act and other regulatory initiatives, electric cooperatives will be operating in an increasingly competitive and deregulated industry. CFC's experience in the telecommunications industry has given it insight into the workings of that environment. An electric distribution system can achieve things a stand-alone company of the same size could not achieve, because it is supported by its G&T, its statewide organization, NRECA and CFC. It can use these resources to differentiate itself from other electric suppliers by bringing economic development to its community or providing diversified services such as natural gas or water and wastewater service. Cooperatives can make themselves even stronger and more able to compete by supporting the organizations that support them. "From the beginning, I have preached that strength comes from unity," Jerome Katzin said in reflecting on CFC's history. "CFC has only one purpose and that is to finance rural electric and telephone systems. It is not in it for the profit. It is not in it for the opportunistic chance that during this season it is profitable or

Sheldon C. Petersen
CFC's Third Governor

Sheldon C. Petersen was working as a staff assistant at Nishnabotna Valley Rural Electric Cooperative in Harlan, Iowa, in 1976 when he was first exposed to CFC. One of his responsibilities was dealing with regulatory issues, and CFC Area Representative Ed Spencer worked with him on a cost-of-service study using the concepts in the Capital Credits Study Committee report.

"I was incredibly impressed," he said. "CFC was a focus for professionalism in the program. It represented the future and growth. It was then that I made the decision that someday I wanted to work for CFC."

In 1980, Petersen became general manager of Rock County Electric Cooperative Association in Janesville, Wisconsin. "That was a wonderful experience for me, and it has really helped me in my work at CFC," he said. "I have been able to understand the needs of the member systems, because I have been behind the manager's desk. I have dealt with the same issues our members deal with every day. As governor, I hope to use that knowledge and understanding to help CFC continue to be responsive to its membership."

Three years after Petersen became manager at Rock County, Spencer announced he was transferring to Arizona. Petersen applied for the job. It was his.

In 1990, Petersen moved to the Washington, D. C., headquarters area to serve as director of policy development and internal audit. He was also named administrative coordinator for the Rural Telephone Finance Cooperative and the National Cooperative Services Corporation. He became director of credit analysis in November 1990.

As one of the staff coordinators for the Member Advisory Committee, he was deeply involved in CFC's strategic planning initiative. He was named corporate secretary in June 1990. He was appointed assistant to the governor in May 1993 and assistant governor and acting administrative officer in June 1994.

As governor, he will have the opportunity to further his objectives for the rural electric program. "We at CFC look forward to continuing to provide the leadership necessary to manage the rural electric program's primary financial resource through this challenging decade of change and beyond," Petersen said.

appetizing to finance a few utilities. It is always going to be there."

The CFC story is an incredible success story. The organization has accomplished the things it was created to do and much, much more. It has enabled its member systems to do together what they could not do individually.

"CFC is a unique financial institution. It is amazing how it was put together 25 years ago, how it has adapted and yet how basically the same it still is," Gill said.

CFC has provided leadership and support for the rural electric program through good times and bad. It has been not a fair-weather friend, but a friend for all seasons.

APPENDIX A

National Rural Electric Cooperative Association
2000 Florida Avenue N.W.
Washington, D.C. 20009

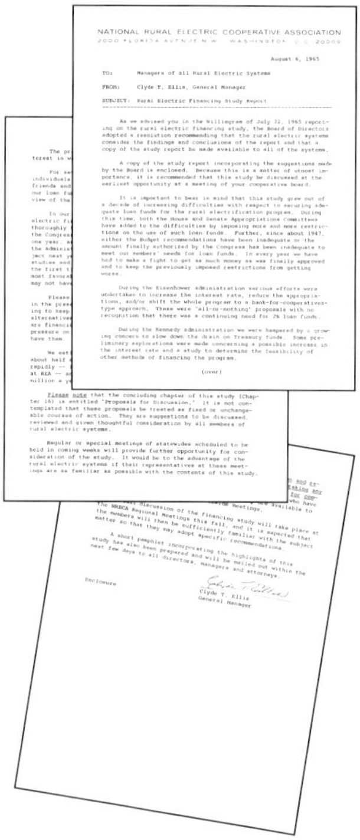

August 6, 1965

To:	Managers of all Rural Electric Systems
From:	Clyde T. Ellis, General Manager
Subject:	Rural Electric Financing Study Report

As we advised you in the Williegram of July 22, 1965 reporting on the rural electric financing study, the Board of Directors adopted a resolution recommending that the rural electric systems consider the findings and conclusions of the report and that a copy of the study report be made available to all of the systems.

A copy of the study report incorporating the suggestions made by the Board is enclosed. Because this is a matter of utmost importance, it is recommended that this study be discussed at the earliest opportunity at a meeting of your cooperative board.

It is important to bear in mind that this study grew out of a decade of increasing difficulties with respect to securing adequate loan funds for the rural electrification program. During this time, both the House and Senate Appropriations Committees have added to the difficulties by imposing more and more restrictions on the use of such loan funds. Further, since about 1947, either the Budget recommendations have been inadequate or the amount finally authorized by the Congress has been inadequate to meet our members' needs for loan funds. In every year we have had to make a fight to get as much money as was finally approved and to keep the previously imposed restrictions from getting worse.

During the Eisenhower administration serious efforts were undertaken to increase the interest rate, reduce the appropriations, and/or shift the whole program to a bank-for-cooperatives-type approach. These were "all-or-nothing" proposals with no recognition that there was a continuing need for 2% loan funds.

During the Kennedy administration we were hampered by a growing concern to slow down the drain on Treasury funds.

Some preliminary explorations were made concerning a possible increase in the interest rate and a study to determine the feasibility of other methods of financing the program. The present Administration has also expressed an active interest in ways and means of reducing the burden on the Treasury.

For several years now, the committees of the Congress and individuals in the Congress, including practically all of our friends and supporters who have had responsibility for obtaining our loan funds, have indicated that something must be done in view of the growing capital needs of the rural electric systems.

In our opinion, knowledge that the NRECA study of rural electric financing was under way has given us time to consider thoroughly the financing problem. Both the Administration and the Congress have been willing to give us this time—at least one year, and possibly two. However, we have been informed by the Administration that there will be legislation on this subject next year. We have been invited to participate in the studies and preparation of that bill. It is my opinion that for the first time in more than a decade we have a Congress which is most favorably inclined toward the rural electric program. We may not have such a favorable Congress after the next election.

Please bear in mind that we are not proposing any change in the present 2% interest rate in the RE Act. We are proposing to keep the law as it is and to add to it other possible alternatives for financing to be utilized by those systems that are financially able to do so. This should help to relieve the pressure on the available 2% loan funds for those systems that have them.

We estimate that we are now getting from the Congress only about half as much funds as we need. The backlog is growing rapidly—it is now over $200 million in applications pending at REA—and it may be expected to reach $300 million to $400 million a year from now.

Please note that the concluding chapter of this study (Chapter 16) is entitled "Proposals for Discussion." It is not contemplated that these proposals be treated as fixed or unchangeable courses of action. They are suggestions to be discussed, reviewed and given thoughtful consideration by all members of rural electric systems.

Regular or special meetings of statewides scheduled to be held in coming weeks will provide further opportunity for consideration of the study. It would be to the advantage of the rural electric systems if their representatives at these meetings are as familiar as possible with the contents of this study.

We urge that this matter be given thoughtful attention and recommend that systems reserve reaching conclusions or taking any policy position until there has been full opportunity for complete discussion of the financing study. We at NRECA who have participated in the development of the study are available to discuss it with you at your statewide meetings.

A full discussion of the financing study will take place at the NRECA Regional Meetings this fall, and it is expected that the members will then be sufficiently familiar with the subject matter so that they may adopt specific recommendations.

A short pamphlet incorporating the highlights of this study has also been prepared and will be mailed out within the next few days to all directors, managers and attorneys.

Clyde T. Ellis
General Manager

Enclosure

A P P E N D I X B **Members of the Board of Directors of CFC**

1971

The First Elected Board of Directors

Vincent P. Slatt
President
District 9, Washington

E. V. Lewis
Vice President
District 2, South Carolina

Charles E. Roemer II
Secretary-Treasurer
District 8, Louisiana

V. Kyle Trout, *District 1, Pennsylvania*

Cecil E. Viverette, *District 1, North Carolina*

I. F. Murph, *District 2, Georgia*

Joe Meng, *District 3, Kentucky*

J. T. Dudley, *District 3, Mississippi*

Arden D. Warner, *District 4, Indiana*

Owen D. Manning, *District 4, Ohio*

Robert R. Wagner, *District 5, Illinois*

Earl L. King, *District 5, Iowa*

Leo E. Grubl, *District 6, South Dakota*

Clarence W. Peterson, *District 6, Minnesota*

Leland Kitt, *District 7, Nebraska*

James A. Golden, *District 7, Colorado*

Allan A. Swanson, *District 8, Missouri*

John L. Rader, *District 9, Alaska*

Paul H. Jones, *District 10, Arizona*

Johnnie M. Ammons, *District 10, Texas*

Louis B. Strong, *District 11, Washington, D.C.*

Robert D. Partridge, *District 11, Washington, D.C.*

1972

Vincent P. Slatt
President
District 9, Washington

E. V. Lewis
Vice President
District 2, South Carolina

Charles E. Roemer II
Secretary-Treasurer
District 8, Louisiana

V. Kyle Trout, *District 1, Pennsylvania*

Cecil E. Viverette, *District 1, North Carolina*

I. F. Murph, *District 2, Georgia*

Joe Meng, *District 3, Kentucky*

J. T. Dudley, *District 3, Mississippi*

Arden D. Warner, *District 4, Indiana*

Owen D. Manning, *District 4, Ohio*

Robert R. Wagner, *District 5, Illinois*

Earl L. King, *District 5, Iowa*

Leo E. Grubl, *District 6, South Dakota*

Clarence W. Peterson, *District 6, Minnesota*

Leland Kitt, *District 7, Nebraska*

James A. Golden, *District 7, Colorado*

Allan A. Swanson, *District 8, Missouri*

John L. Rader, *District 9, Alaska*

Paul H. Jones, *District 10, Arizona*

Johnnie M. Ammons, *District 10, Texas*

Louis B. Strong, *District 11, Washington, D.C.*

Robert D. Partridge, *District 11, Washington, D.C.*

1973

Vincent P. Slatt
President
District 9, Washington

E. V. Lewis
Vice President
District 2, South Carolina

John L. Rader
Secretary-Treasurer
District 9, Alaska

V. Kyle Trout, *District 1, Pennsylvania*

Cecil E. Viverette, *District 1, North Carolina*

I. F. Murph, *District 2, Georgia*

Joe Meng, *District 3, Kentucky*

J. T. Dudley, *District 3, Mississippi*

Arden D. Warner, *District 4, Indiana*

Owen D. Manning, *District 4, Ohio*

Robert R. Wagner, *District 5, Illinois*

Earl L. King, *District 5, Iowa*

Leo E. Grubl, *District 6, South Dakota*

Clarence W. Peterson, *District 6, Minnesota*

Leland Kitt, *District 7, Nebraska*

James A. Golden, *District 7, Colorado*

Allan A. Swanson, *District 8, Missouri*

Charles E. Roemer II, *District 8, Louisiana*

Paul H. Jones, *District 10, Arizona*

Johnnie M. Ammons, *District 10, Texas*

Charles E. Wyckoff, *District 11, Washington, D.C.*

Robert D. Partridge, *District 11, Washington, D.C.*

1974

Vincent P. Slatt,
President
District 9, Washington

E. V. Lewis,
Vice President
District 2, South Carolina

John L. Rader
Secretary-Treasurer
District 9, Alaska

V. Kyle Trout, *District 1, Pennsylvania*

Cecil E. Viverette, *District 1, North Carolina*

I. F. Murph, *District 2, Georgia*

Joe Meng, *District 3, Kentucky*

J. T. Dudley, *District 3, Mississippi*

Arden D. Warner, *District 4, Indiana*

Owen D. Manning, *District 4, Ohio*

Robert R. Wagner, *District 5, Illinois*

Earl L. King, *District 5, Iowa*

Leo E. Grubl, *District 6, South Dakota*

Clarence W. Peterson, *District 6, Minnesota*

Leland Kitt, *District 7, Nebraska*

James A. Golden, *District 7, Colorado*

Allan A. Swanson, *District 8, Missouri*

Lloyd C. Patton, *District 8, Oklahoma*

Paul H. Jones, *District 10, Arizona*

Johnnie M. Ammons, *District 10, Texas*

Charles E. Wyckoff, *District 11, Washington, D.C.*

Robert D. Partridge, *District 11, Washington, D.C.*

1975

Vincent P. Slatt
President
District 9, Washington

E. V. Lewis
Vice President
District 2, South Carolina

John L. Rader
Secretary-Treasurer
District 9, Alaska

V. Kyle Trout, *District 1, Pennsylvania*

Cecil E. Viverette, *District 1, North Carolina*

I. F. Murph, *District 2, Georgia*

Joe Meng, *District 3, Kentucky*

J. T. Dudley, *District 3, Mississippi*

Arden D. Warner, *District 4, Indiana*

Owen D. Manning, *District 4, Ohio*

Robert R. Wagner, *District 5, Illinois*

Earl L. King, *District 5, Iowa*

Leo E. Grubl, *District 6, South Dakota*

Clarence W. Peterson, *District 6, Minnesota*

Leland Kitt, *District 7, Nebraska*

James A. Golden, *District 7, Colorado*

Allan A. Swanson, *District 8, Missouri*

Lloyd C. Patton, *District 8, Oklahoma*

Paul H. Jones, *District 10, Arizona*

Johnnie M. Ammons, *District 10, Texas*

John R. Dolinger, *District 11, Washington, D.C.*

Robert D. Partridge, *District 11, Washington, D.C.*

1976

Vincent P. Slatt
President
District 9, Washington

J. T. Dudley
Vice President
District 3, Mississippi

John L. Rader
Secretary-Treasurer
District 9, Alaska

V. Kyle Trout, *District 1, Pennsylvania*

Cecil E. Viverette, *District 1, North Carolina*

I. F. Murph, *District 2, Georgia*

Leon E. Weaver, *District 2, Florida*

Joe Meng, *District 3, Kentucky*

Arden D. Warner, *District 4, Indiana*

Owen D. Manning, *District 4, Ohio*

Robert R. Wagner, *District 5, Illinois*

Earl L. King, *District 5, Iowa*

Leo E. Grubl, *District 6, South Dakota*

Clarence W. Peterson, *District 6, Minnesota*

Leland Kitt, *District 7, Nebraska*

James A. Golden, *District 7, Colorado*

Allan A. Swanson, *District 8, Missouri*

Lloyd C. Patton, *District 8, Oklahoma*

Paul H. Jones, *District 10, Arizona*

Johnnie M. Ammons, *District 10, Texas*

John R. Dolinger, *District 11, Washington, D.C.*

Robert D. Partridge, *District 11, Washington, D.C.*

1977

J. T. Dudley
President
District 3, Mississippi

Owen D. Manning
Vice President
District 4, Ohio

Clarence W. Peterson
Secretary-Treasurer
District 6, Minnesota

Frank W. Blake, *District 1, Virginia*

Cecil E. Viverette, *District 1, North Carolina*

Roy L. Sanders, *District 2, Georgia*

Leon E. Weaver, *District 2, Florida*

Joe Meng, *District 3, Kentucky*

Donald R. Norris, *District 5, Iowa*

Robert R. Wagner, *District 5, Illinois*

Leo E. Grubl, *District 6, South Dakota*

Merl D. Rissler, *District 7, Wyoming*

Jack D. Goodman, *District 7, Kansas*

Allan A. Swanson, *District 8, Missouri*

Jack L. Williams, *District 8, Arkansas*

John L. Rader, *District 9, Alaska*

Clement W. Eaton, *District 9, Idaho*

Paul H. Jones, *District 10, Arizona*

Don J. Hart, *District 10, Texas*

Gene W. Porter, *District 11, Washington, D.C.*

Robert D. Partridge, *District 11, Washington, D.C.*

1978

Owen D. Manning
President
District 4, Ohio

Donald R. Norris
Vice President
District 5, Iowa

Clarence W. Peterson
Secretary-Treasurer
District 6, Minnesota

Frank W. Blake, *District 1, Virginia*

Robert L. Butts, *District 1, Pennsylvania*

Roy L. Sanders, *District 2, Georgia*

Leon E. Weaver, *District 2, Florida*

Joe Meng, *District 3, Kentucky*

James A. Vann, Jr., *District 3, Alabama*

C. Frank Swanson, *District 4, Indiana*

Robert R. Wagner, *District 5, Illinois*

Loren Paulsen, *District 6, South Dakota*

Merl D. Rissler, *District 7, Wyoming*

C. J. Hoke, *District 7, Nebraska*

Allan A. Swanson, *District 8, Missouri*

Jack L. Williams, *District 8, Arkansas*

Wellen A. Jones, *District 9, Montana*

Clement W. Eaton, *District 9, Idaho*

Edward L. Anway, District 10, Arizona

Don J. Hart, *District 10, Texas*

Gene W. Porter, *District 11, Washington, D.C.*

Robert D. Partridge, *District 11, Washington, D.C.*

1979

Donald R. Norris
President
District 5, Iowa

Leon E. Weaver
Vice President
District 2, Florida

Jack L. Williams
Secretary-Treasurer
District 8, Arkansas

Frank W. Blake, *District 1, Virginia*

Robert L. Butts, *District 1, Pennsylvania*

Roy L. Sanders, *District 2, Georgia*

James A. Vann, Jr., *District 3, Alabama*

Walter S. Nunnelly, *District 3, Tennessee*

George Charles Manning, *District 4, Ohio*

C. Frank Swanson, *District 4, Indiana*

Allen D. Beadles, *District 5, Wisconsin*

Loren Paulsen, *District 6, South Dakota*

Meridith A. Haslerud, *District 6, Minnesota*

Merl D. Rissler, *District 7, Wyoming*

C. J. Hoke, *District 7, Nebraska*

B. Dean Sanger, *District 8, Oklahoma*

Wellen A. Jones, *District 9, Montana*

Clement W. Eaton, *District 9, Idaho*

Edward L. Anway, *District 10, Arizona*

Don J. Hart, *District 10, Texas*

D. L. Knight, *District 11, Washington, D.C.*

Robert D. Partridge, *District 11, Washington, D.C.*

1980

Leon E. Weaver
President
District 2, Florida

Frank W. Blake
Vice President
District 1, Virginia

Jack L. Williams
Secretary-Treasurer
District 8, Arkansas

Robert L. Butts, *District 1, Pennsylvania*

Roy L. Sanders, *District 2, Georgia*

James A. Vann, Jr., *District 3, Alabama*

Walter S. Nunnelly, *District 3, Tennessee*

George Charles Manning, *District 4, Ohio*

C. Frank Swanson, *District 4, Indiana*

Allen D. Beadles, *District 5, Wisconsin*

C. E. Ferguson, *District 5, Illinois*

Loren Paulsen, *District 6, South Dakota*

Meridith A. Haslerud, *District 6, Minnesota*

Merl D. Rissler, *District 7, Wyoming*

C. J. Hoke, *District 7, Nebraska*

B. Dean Sanger, *District 8, Oklahoma*

Wellen A. Jones, *District 9, Montana*

Clement W. Eaton, *District 9, Idaho*

Edward L. Anway, *District 10, Arizona*

Don J. Hart, *District 10, Texas*

D. L. Knight, *District 11, Washington, D.C.*

Robert D. Partridge, *District 11, Washington, D.C.*

1981

Leon E. Weaver
President
District 2, Florida

Frank W. Blake
Vice President
District 1, Virginia

Jack L. Williams
Secretary-Treasurer
District 8, Arkansas

Robert L. Butts, *District 1, Pennsylvania*

Roy L. Sanders, *District 2, Georgia*

James A. Vann, Jr., *District 3, Alabama*

Walter S. Nunnelly, *District 3, Tennessee*

George Charles Manning, *District 4, Ohio*

C. Frank Swanson, *District 4, Indiana*

Allen D. Beadles, *District 5, Wisconsin*

C. E. Ferguson, *District 5, Illinois*

Loren Paulsen, *District 6, South Dakota*

Meridith A. Haslerud, *District 6, Minnesota*

Dean Reese, *District 7, Kansas*

C. J. Hoke, *District 7, Nebraska*

B. Dean Sanger, *District 8, Oklahoma*

Wellen A. Jones, *District 9, Montana*

Clement W. Eaton, *District 9, Idaho*

Edward L. Anway, *District 10, Arizona*

Don J. Hart, *District 10, Texas*

Angus S. Hastings, *District 11, Washington, D.C.*

Robert D. Partridge, *District 11, Washington, D.C.*

1982

Jack L. Williams
President
District 8, Arkansas

Frank W. Blake
Vice President
District 1, Virginia

C. J. Hoke
Secretary-Treasurer
District 7, Nebraska

Robert L. Butts, *District 1, Pennsylvania*

John C. Anderson, *District 2, South Carolina*

Roy L. Sanders, *District 2, Georgia*

James A. Vann, Jr., *District 3, Alabama*

Walter S. Nunnelly, *District 3, Tennessee*

George Charles Manning, *District 4, Ohio*

C. Frank Swanson, *District 4, Indiana*

Allen D. Beadles, *District 5, Wisconsin*

C. E. Ferguson, *District 5, Illinois*

Loren Paulsen, *District 6, South Dakota*

Meridith A. Haslerud, *District 6, Minnesota*

Dean Reese, *District 7, Kansas*

B. Dean Sanger, *District 8, Oklahoma*

Wellen A. Jones, *District 9, Montana*

Clement W. Eaton, *District 9, Idaho*

Edward L. Anway, *District 10, Arizona*

Don J. Hart, *District 10, Texas*

Angus S. Hastings, *District 11, Washington, D.C.*

Robert D. Partridge, *District 11, Washington, D.C.*

1983

James A. Vann, Jr.
President
District 3, Alabama

C. J. Hoke
Vice President
District 7, Nebraska

Robert L. Butts
Secretary-Treasurer
District 1, Pennsylvania

James H. Garner, *District 1, North Carolina*

William H. Newberry, Sr., *District 2, Georgia*

M. Larry Longshore, *District 2, South Carolina*

Walter S. Nunnelly, *District 3, Tennessee*

George Charles Manning, *District 4, Ohio*

C. Frank Swanson, *District 4, Indiana*

Allen D. Beadles, *District 5, Wisconsin*

C. E. Ferguson, *District 5, Illinois*

Loren Paulsen, *District 6, South Dakota*

Meridith A. Haslerud, *District 6, Minnesota*

Dean Reese, *District 7, Kansas*

B. Dean Sanger, *District 8, Oklahoma*

E. W. Patrick, *District 8, Louisiana*

Wellen A. Jones, *District 9, Montana*

James F. Palin, *District 9, Alaska*

Edward L. Anway, *District 10, Arizona*

Alton Higginbotham, *District 10, Texas*

Guy C. Lewis, Jr., *District 11, Washington, D.C.*

Robert D. Partridge, *District 11, Washington, D.C.*

1984

B. Dean Sanger
President
District 8, Oklahoma

Wellen A. Jones
Vice President
District 9, Montana

George Charles Manning
Secretary-Treasurer
District 4, Ohio

James H. Garner, *District 1, North Carolina*

James M. Reynolds, *District 1, Virginia*

William H. Newberry, Sr., *District 2, Georgia*

M. Larry Longshore, *District 2, South Carolina*

Hollis H. Alford, *District 3, Mississippi*

Thomas Barker, Jr., *District 3, Kentucky*

Jimmy M. Gebert, *District 4, Indiana*

Allen D. Beadles, *District 5, Wisconsin*

C. E. Ferguson, *District 5, Illinois*

Leonard C. Welter, *District 6, South Dakota*

Meridith A. Haslerud, *District 6, Minnesota*

Dean Reese, *District 7, Kansas*

Richard L. Arnold, *District 7, Colorado*

E. W. Patrick, *District 8, Louisiana*

James F. Palin, *District 9, Alaska*

Lyn R. Borah, *District 10, Arizona*

Alton Higginbotham, *District 10, Texas*

Guy C. Lewis, Jr., *District 11, Washington, D.C.*

Bob Bergland, *District 11, Washington, D.C.*

1985

Alton Higginbotham
President
District 10, Texas

Dean Reese
Vice President
District 7, Kansas

C. E. Ferguson
Secretary-Treasurer
District 5, Illinois

James H. Garner, *District 1, North Carolina*

James M. Reynolds, *District 1, Virginia*

William H. Newberry, Sr., *District 2, Georgia*

M. Larry Longshore, *District 2, South Carolina*

Hollis H. Alford, *District 3, Mississippi*

Thomas Barker, Jr., *District 3, Kentucky*

Jimmy M. Gebert, *District 4, Indiana*

Donald L. Grimes, *District 4, Michigan*

Donald E. Applegate, *District 5, Iowa*

Leonard C. Welter, *District 6, South Dakota*

Philip O. Martin, *District 6, Minnesota*

Richard L. Arnold, *District 7, Colorado*

E. W. Patrick, *District 8, Louisiana*

Robert M. Alderson, *District 8, Missouri*

Ronald C. Ostberg, *District 9, Montana*

James F. Palin, *District 9, Alaska*

Lyn R. Borah, *District 10, Arizona*

Vernon C. Williams, *District 11, Washington, D.C.*

Bob Bergland, *District 11, Washington, D.C.*

1986

Alton Higginbotham
President
District 10, Texas

Dean Reese
Vice President
District 7, Kansas

M. Larry Longshore
Secretary-Treasurer
District 2, South Carolina

James H. Garner, *District 1, North Carolina*

James M. Reynolds, *District 1, Virginia*

William H. Newberry, Sr., *District 2, Georgia*

Hollis H. Alford, *District 3, Mississippi*

Thomas Barker, Jr., *District 3, Kentucky*

Joseph H. Huber, *District 4, Indiana*

Donald I. Grimes, *District 4, Michigan*

Donald E. Applegate, *District 5, Iowa*

Fred J. Harasha, *District 5, Wisconsin*

Leonard C. Welter, *District 6, South Dakota*

Philip O. Martin, *District 6, Minnesota*

Richard L. Arnold, *District 7, Colorado*

E. W. Patrick, *District 8, Louisiana*

Robert M. Alderson, *District 8, Missouri*

Ronald C. Ostberg, *District 9, Montana*

James F. Palin, *District 9, Alaska*

Lyn R. Borah, *District 10, Arizona*

Vernon C. Williams, *District 11, Washington, D.C.*

Bob Bergland, *District 11, Washington, D.C.*

1987

James H. Garner
President
District 1, North Carolina

James F. Palin
Vice President
District 9, Alaska

M. Larry Longshore
Secretary-Treasurer
District 2, South Carolina

James M. Reynolds, *District 1, Virginia*

William H. Newberry, Sr., *District 2, Georgia*

Hollis H. Alford, *District 3, Mississippi*

Thomas Barker, Jr., *District 3, Kentucky*

Joseph H. Huber, *District 4, Indiana*

Donald I. Grimes, *District 4, Michigan*

Donald E. Applegate, *District 5, Iowa*

Fred J. Harasha, *District 5, Wisconsin*

Leonard C. Welter, *District 6, South Dakota*

Philip O. Martin, *District 6, Minnesota*

Dean Reese, *District 7, Kansas*

Ronald R. Harper, *District 7, Wyoming*

E. W. Patrick, *District 8, Louisiana*

Robert M. Alderson, *District 8, Missouri*

Ronald C. Ostberg, *District 9, Montana*

Richard J. Marquez, *District 10, New Mexico*

Don J. Hart, *District 10, Texas*

Don M. Heathington, *District 11, Washington, D.C.*

Bob Bergland, *District 11, Washington, D.C.*

1988

James H. Garner
President
District 1, North Carolina

James F. Palin
Vice President
District 9, Alaska

Thomas Barker, Jr.
Secretary-Treasurer
District 3, Kentucky

James M. Reynolds, *District 1, Virginia*

William H. Newberry, Sr., *District 2, Georgia*

J. E. (Gene) Smith, *District 2, Florida*

Hollis H. Alford, *District 3, Mississippi*

Joseph H. Huber, *District 4, Indiana*

John M. (Mike) McBride, *District 4, Ohio*

Donald E. Applegate, *District 5, Iowa*

Fred J. Harasha, *District 5, Wisconsin*

Leonard C. Welter, *District 6, South Dakota*

Philip O. Martin, *District 6, Minnesota*

Dean Reese, *District 7, Kansas*

Ronald R. Harper, *District 7, Wyoming*

E. W. Patrick, *District 8, Louisiana*

Robert M. Alderson, *District 8, Missouri*

Ronald C. Ostberg, *District 9, Montana*

Richard J. Marquez, *District 10, New Mexico*

Bill McGinnis, *District 10, Texas*

Don M. Heathington, *District 11, Washington, D.C.*

Bob Bergland, *District 11, Washington, D.C.*

1989

Hollis H. Alford
President
District 3, Mississippi

Thomas Barker, Jr
Vice President
District 3, Kentucky

Philip O. Martin
Secretary-Treasurer
District 6, Minnesota

Johnnie R. Austin, *District 1, North Carolina*

James M. Reynolds, *District 1, Virginia*

John B. Floyd, Jr., *District 2, Georgia*

J. E. (Gene) Smith, *District 2, Florida*

Joseph H. Huber, *District 4, Indiana*

John M. (Mike) McBride, *District 4, Ohio*

Donald E. Applegate, *District 5, Iowa*

Fred J. Harasha, *District 5, Wisconsin*

Leonard C. Welter, *District 6, South Dakota*

William R. Haight, *District 7, Colorado*

Ronald R. Harper, *District 7, Wyoming*

Robert M. Alderson, *District 8, Missouri*

Tom Sloan, *District 8, Arkansas*

Ronald C. Ostberg, *District 9, Montana*

David E. Piper, *District 9, Oregon*

Richard J. Marquez, *District 10, New Mexico*

Bill McGinnis, *District 10, Texas*

John W. (Jack) Williams, Jr., *District 11, Washington, D.C.*

Bob Bergland, *District 11, Washington, D.C.*

1990

Hollis H. Alford
President
District 3, Mississippi

Philip O. Martin
Vice President
District 6, Minnesota

Donald E. Applegate
Secretary-Treasurer
District 5, Iowa

John C. Anderson, *District 1, Virginia*

Johnnie R. Austin, *District 1, North Carolina*

John B. Floyd, Jr., *District 2, Georgia*

J. E. (Gene) Smith, *District 2, Florida*

Ralph L. Loveless, *District 3, Tennessee*

Joseph H. Huber, *District 4, Indiana*

John M. (Mike) McBride, *District 4, Ohio*

Fred J. Harasha, *District 5, Wisconsin*

Bill Bertram, *District 6, North Dakota*

William R. Haight, *District 7, Colorado*

Ronald R. Harper, *District 7, Wyoming*

Robert M. Alderson, *District 8, Missouri*

Tom Sloan, *District 8, Arkansas*

Ronald C. Ostberg, *District 9, Montana*

David E. Piper, *District 9, Oregon*

Richard J. Marquez, *District 10, New Mexico*

Bill McGinnis, *District 10, Texas*

John W. (Jack) Williams, Jr., *District 11, Washington, D.C.*

Bob Bergland, *District 11, Washington, D.C.*

1991

J. E. (Gene) Smith
President
District 2, Florida

John M. (Mike) McBride
Vice President
District 4, Ohio

Richard J. Marquez
Secretary-Treasurer
District 10, New Mexico

John C. Anderson, *District 1, Virginia*

Johnnie R. Austin, *District 1, North Carolina*

John B. Floyd, Jr., *District 2, Georgia*

Ralph L. Loveless, *District 3, Tennessee*

Terry Pitchford, *District 3, Alabama*

Joseph H. Huber, *District 4, Indiana*

Harold I. Dycus, *District 5, Illinois*

Fred J. Harasha, *District 5, Wisconsin*

Bill Bertram, *District 6, North Dakota*

Garry O. Bye, *District 6, Minnesota*

William R. Haight, *District 7, Colorado*

Ronald R. Harper, *District 7, Wyoming*

J. Chris Cariker, *District 8, Oklahoma*

Tom Sloan, *District 8, Arkansas*

Gordon J. Hudson, *District 9, Idaho*

David E. Piper, *District 9, Oregon*

Bill McGinnis, *District 10, Texas*

Jay A. York, *District 11, Washington, D.C.*

Bob Bergland, *District 11, Washington, D.C.*

1992

J. E. (Gene) Smith
President
District 2, Florida

John M. (Mike) McBride
Vice President
District 4, Ohio

Richard J. Marquez
Secretary-Treasurer
District 10, New Mexico

John C. Anderson, *District 1, Virginia*

Johnnie R. Austin, *District 1, North Carolina*

John B. Floyd, Jr., *District 2, Georgia*

Ralph L. Loveless, *District 3, Tennessee*

Terry Pitchford, *District 3, Alabama*

Joseph H. Huber, *District 4, Indiana*

Robert J. Bauman, *District 5, Iowa*

Harold I. Dycus, *District 5, Illinois*

Bill Bertram, *District 6, North Dakota*

Garry O. Bye, *District 6, Minnesota*

Nadine Griffin, *District 7, Kansas*

Ronald R. Harper, *District 7, Wyoming*

J. Chris Cariker, *District 8, Oklahoma*

Tom Sloan, *District 8, Arkansas*

Gordon J. Hudson, *District 9, Idaho*

David E. Piper, *District 9, Oregon*

Bill McGinnis, *District 10, Texas*

Jay A. York, *District 11, Washington, D.C.*

Bob Bergland, *District 11, Washington, D.C.*

1993

Bill McGinnis
President
District 10, Texas

John M. (Mike) McBride
Vice President
District 4, Ohio

Johnnie R. Austin
Secretary-Treasurer
District 1, North Carolina

John C. Anderson, *District 1, Virginia*

John B. Floyd, Jr., *District 2, Georgia*

J. E. (Gene) Smith, *District 2, Florida*

Ralph L. Loveless, *District 3, Tennessee*

Terry Pitchford, *District 3, Alabama*

Ralph Harmeyer, *District 4, Indiana*

Robert J. Bauman, *District 5, Iowa*

Harold I. Dycus, *District 5, Illinois*

Bill Bertram, *District 6, North Dakota*

Garry O. Bye, *District 6, Minnesota*

Nadine Griffin, *District 7, Kansas*

Paul J. Liess, *District 7, Nebraska*

J. Chris Cariker, *District 8, Oklahoma*

Tom Sloan, *District 8, Arkansas*

Gordon J. Hudson, *District 9, Idaho*

David E. Piper, *District 9, Oregon*

George W. Kline, *District 10, Arizona*

J. C. (Jim) Roberts, *District 11, Washington, D.C.*

Bob Bergland, *District 11, Washington, D.C.*

1994

Bill McGinnis
President
District 10, Texas

J. Chris Cariker
Vice President
District 8, Oklahoma

Garry O. Bye
Secretary-Treasurer
District 6, Minnesota

John C. Anderson, *District 1, Virginia*

Johnnie R. Austin, *District 1, North Carolina*

John B. Floyd, Jr., *District 2, Georgia*

Robert O. Williams, *District 2, South Carolina*

Ralph L. Loveless, *District 3, Tennessee*

Terry Pitchford, *District 3, Alabama*

Ralph Harmeyer, *District 4, Indiana*

Gerard P. Paolucci, *District 4, Ohio*

Robert J. Bauman, *District 5, Iowa*

Harold I. Dycus, *District 5, Illinois*

Bill Bertram, *District 6, North Dakota*

Nadine Griffin, *District 7, Kansas*

Paul J. Liess, *District 7, Nebraska*

Tom Sloan, *District 8, Arkansas*

Gordon J. Hudson, *District 9, Idaho*

David E. Piper, *District 9, Oregon*

George W. Kline, *District 10, Arizona*

J. C. (Jim) Roberts, *District 11, Washington, D.C.*

Glenn English, *District 11, Washington, D.C.*

APPENDIX C

Who's Who in CFC History

Truman L. Brandt served as CFC's chief financial officer from February 1984 until his retirement in 1993.

Richard B. Bulman joined CFC as power supply officer in March 1980. He was appointed loan officer in June 1984 and senior vice president and loan officer in June 1992. Before joining CFC, he served as deputy administrator of the Southwestern Power Administration and in various power supply positions with REA.

Milton A. Chase served as a consultant to CFC during its formative period and was appointed engineering officer in March 1972, a position he held until his retirement in May 1980. He joined the REA in 1937 and served in the Department of Interior during the Kennedy and Johnson administrations. He became a consultant in 1969 and participated in the establishment of the New England Power Pool.

Norman L. Clapp served as administrator of REA from 1961 to 1969 and was an active participant in the work of the Long-Range Study Committee.

J. R. Cobb, manager of Texas Electric Cooperatives, Inc., served as chairman of the Implementation Subcommittee of the Long-Range Study Committee. He was a recipient of CFC's Pioneer Award.

Martin R. Crowson joined CFC as a senior financial consultant in June 1982 and became the director of member finance in June 1993.

Clyde Denton joined CFC in January 1971 as information and member relations officer. He was appointed special assistant to the governor in June 1987, a post he held until his retirement in 1989. Before his work with CFC, he was manager of information services for the Kentucky Association of Rural Electric Cooperatives.

Clyde T. Ellis served as general manager of NRECA from 1942 until 1968. He fostered the concept of supplemental financing and continued to serve as an advisor to the Long-Range Study Committee after his retirement from NRECA.

Paul Fox was CFC's first finance officer and served in that position from September 1970 until his retirement in May 1983.

Charles B. Gill served as CFC's borrowers operations officer from April 1972 until he was appointed deputy governor in August 1978. He became governor of CFC in September 1979. Before joining CFC, he held various positions with REA.

Gary A. Gordy served CFC in many capacities, including service coordinator, 1975–77; director of borrowers services, 1977–78; borrowers operations officer, 1978–80; loan officer, 1980–84; planning and member relations officer, 1984–89; and assistant to the governor from January 1989 until his retirement in 1993. Before joining CFC, he was vice president of marketing for United Utility Supply, Inc., and general manager of White County Rural Electric Membership Corporation.

David A. Hamil served as administrator of REA from 1956 to 1961 and during CFC's formative years from 1969 to 1978. He helped to organize and served as a director of Highline Electric Association in Colorado. He received CFC's Pioneer Award.

David J. Hedberg joined CFC as director of rate design and analysis in December 1981 and serves as director of rates and special projects. Before joining CFC, he held the position of public utilities specialist–industry economist with the Federal Energy Regulatory Commission.

Virgil Herriott was general manager of the Sioux Valley Empire Electric Association in Colman, South Dakota. He served as chairman of the Viewpoints and Objectives Subcommittee of the Long-Range Study Committee. He was an incorporator of CFC and a member of its first board of directors. He also served as chairman of the Viewpoints and Objectives Subcommittee of the CFC Committee on Objectives and Planning. He was a recipient of CFC's Pioneer Award.

Jerome S. Katzin, a partner in Kuhn, Loeb & Company and managing director of its successor, Lehman Brothers Kuhn Loeb, Inc., served as a consultant to the Long-Range Study Committee and is credited with originating the basic concept of CFC. He served as CFC's financial advisor from its incorporation in 1969 until 1989. He received CFC's Pioneer Award.

Leland Leatherman served as attorney for the Arkansas Electric Cooperatives, Inc., and other Arkansas cooperatives. He was a member of the Long-Range Study Committee and an incoporator of CFC, for which he received CFC's Pioneer award. He also served as CFC's first secretary-treasurer.

Steven L. Lilly joined CFC as a senior financial consultant in 1983. He became director of special finance in June 1985, director of corporate finance in June 1986 and treasurer and principal finance officer in June 1993. He was appointed senior vice president and chief financial officer in January 1994.

John Jay List joined CFC as staff attorney in February 1972. He subsequently held the positions of assistant corporate counsel and associate corporate counsel. He served as corporate counsel from 1980 until 1991 when he was appointed general counsel. He was appointed senior vice president and general counsel in June 1992.

Owen Manning became general manager of The Frontier Power Company in Coshocton, Ohio, in 1955, a position he held until his death in 1982. He was a member of the Long-Range Study Committee and served CFC as vice president and president. He received CFC's Pioneer Award.

Roger Oresman was a partner with Milbank, Tweed, Hadley & McCloy and directed that firm's work as counsel to CFC. He received CFC's Pioneer Award.

Robert D. Partridge served as executive vice president and general manager of NRECA from 1968 until his retirement in April 1984. Before that, he held various positions with NRECA and REA. He was actively involved in the work of the Long-Range Study Committee. He was an incorporator of CFC and served on its board of directors from its inception until 1984. He received CFC's Pioneer Award.

Sheldon C. Petersen joined CFC as an area representative in July 1983. He became director of policy development and internal audit in January 1990, director of credit analysis in November 1990 and corporate secretary in June 1992. He was appointed assistant to the governor in May 1993. He became assistant governor and acting administrative officer in June 1994. On December 5, 1994, the board selected him as the third governor and chief executive officer of CFC.

Louis A. Roland joined CFC as associate corporate counsel in June 1971. He was appointed corporate counsel in June 1972, a position he held until his retirement in July 1979. Before his work with CFC, Roland held various positions in the Office of the General Counsel of the Department of Agriculture, including deputy director of the Rural Electrification Division.

Angelo M. Salera joined CFC's staff as controller in September 1975.

Charles U. Samenow served as CFC's corporate counsel from its inception until his death in 1972. He joined REA in 1940 and subsequently served as interagency and legislative consultant to the administrator until his retirement in 1968. He served as a special consultant to the Long-Range Study Committee. He was a recipient of CFC's Pioneer Award.

Ira Shesser was CFC's first loan officer, holding that position from September 1970 until June 1980, when he was appointed assistant to the governor. He served in that position until his retirement in May 1985. He was a recipient of CFC's Pioneer Award. Before his work with CFC, he was legislative research coordinator for NRECA and served on the legislative staff of REA.

Vincent P. Slatt was manager of Inland Power and Light Company in Spokane, Washington. He served as chairman of the Financing Subcommittee of the Long-Range Study Committee. He was an incorporator of CFC, a member of its first board of directors and president of its first elected board of directors. He was a recipient of CFC's Pioneer Award.

J. K. Smith was the first governor of CFC, a position he held from November 1970 until his retirement in September 1979. He was also the first president of CFC. He served as chairman of the Long-Range Study Committee. Before his work with CFC, he served as executive manager of the Kentucky Rural Electric Cooperative Corporation and manager of Fleming-Mason Rural Electric Cooperative Corporation.

Nicholas J. Walsh joined CFC as administrative officer in August 1971, a position he held until his retirement in May 1994.

GLOSSARY

Administrator, REA

In the rural electric industry, the administrator is the chief executive officer of the Rural Electrification Administration (REA).

All-requirements power contract

A contract primarily used by municipal electric and rural electric systems that describes a formal agreement entered into by a power supply system and its member distribution systems. In this contract, the distribution systems agree to purchase all their wholesale power needs from the power supply system at rates prescribed in the agreement and adjusted periodically to meet the power supply system's cost of providing the power.

Blue sky laws

Various states have enacted blue sky laws to regulate the sale and issuance of securities. These laws are designed to prevent fraud in the sale and disposition of securities.

Bond

A certificate stating that you have lent a company money. According to this agreement, in return for your money, the company promises to pay you a certain amount of money based on the certificate's interest rate, usually at specific intervals. The company also agrees to repay you the full amount of your loan at a stated date (maturity), usually from 1 to 30 years. A bond is a debt, rather than an equity, investment. Stocks of CFC's Capital Term Certificates are considered equity investments.

Guaranteed bond

Bonds on which the principal or interest, or both, is guaranteed by another corporation or a parent company in case of default by the issuing corporation.

Mortgage bonds

Bonds whose security (support) is specific property owned by the corporation. The bond is issued in much the same way as a house mortgage is secured by a home. The investor is comforted by the knowledge that if the corporation becomes unable to make the interest payments, the property can be sold to pay the bondholder.

Bond anticipation notes (BAN)

Short-term notes sold in anticipation of a bond issue and then retired by proceeds from the sale of the bonds.

Capital credits

The amounts of a cooperative's net margins allocated to individual patrons. The allocation is usually based on the patron's use of the cooperative's service. CFC's member patronage is determined on the basis of the total interest a member pays on outstanding CFC debt during a fiscal year. There are numerous methods of capital credits retirement. The most commonly used are last in, first out (LIFO); first in, first out (FIFO); and the percentage method.

Capital Term Certificates (CTCs)

CTCs represent debt of CFC that is subordinated to all other indebtedness, including that incurred in the capital markets through the sale of CFC bonds and commercial paper and through bank loans. Unsecured, subordinated investments in CFC to provide equity for loans or other equity requirements.

Collateral trust bonds

Bonds secured by the pledge of collateral with a trustee under a collateral trust indenture. CFC raises most of its long-term capital by marketing collateral trust bonds in the private capital markets, backed by the mortgages of its member systems.

Commercial paper

Short-term investments that have a definite maturity date. CFC sells commercial paper to members and investors with maturities ranging from 3 to 270 days. CFC's commercial paper pays investors money market rates and serves as an excellent way to earn a return on funds not in immediate use.

Common mortgage, REA-CFC

A mortgage held in common by REA and CFC. It is used to secure loans made concurrently by the two agencies. The mortgage provides the CFC lien an equal proratable status with the REA lien.

Concurrent loan

A CFC long-term loan made concurrently with an REA long-term loan. Plant revenue ratio, a relative measurement of the productivity of a plant, determines the amount of supplemental capital each borrower will receive. Based on this ratio, REA provides 90, 80 or 70 percent of a borrower's long-term needs. CFC supplies the remaining 10, 20 or 30 percent supplemental portion of the loan. Systems with concurrent loans draw down funds in the following sequence:

1. Requisition REA for the first half of its portion of the loan.

2. Requisition CFC for the entire supplemental portion.
3. Draw down the remaining REA funds.

Credit rating

An estimate of the creditworthiness of an organization involved in the sale of securities by a credit servicing organization. Moody's Investors Services and Standards and Poor's Corporation are examples of such organizations.

Debt service coverage (DSC)

A financial measurement relating to the borrower's ability to generate sufficient funds to cover the cash requirements of its total indebtedness (principal and interest) on an annual basis. For a system's mortgage to be eligible collateral for CFC's bonds, the system must have an average DSC of 1.25 of the two highest of the last three years, which means that the borrower must have annual margins equal to 1.25 times its total annual debt obligations. A recommended DSC level would need to be substantially higher than the minimum level.

Debt-to-equity ratio

The amount of a system's indebtedness in relation to its equity. A measurement of the relationship between total debt of an organization and the net ownership of plant.

Direct, or private, placement

The direct sale of a new security by the issuer to investors, bypassing the underwriter or middleman.

Equity

The ownership right or the risk interest in property. It can also be described as the monetary value of a property or business that exceeds the claims and/or liens against it by others. In a cooperative, equity is the value of member ownership plus financial support.

Equity management

A set of financial goals and objectives that a rural electric co-op adopts to establish a prudent blending of debt costs, capital credits cycles, TIER (debt coverage) levels, and equity targets.

Federal Financing Bank (FFB)

The federal agency that finances federal programs by purchasing loan assets from agencies, making direct loans for various agencies and purchasing agency debt securities. It then issues government securities in the public markets or it borrows from the U.S. Treasury. The REA's guarantee program is funded by federal security sales made by the FFB.

First in, first out (FIFO)

In a cooperative structure, this term refers to a method of retiring a member's capital credits. Those capital credits accrued to the cooperative in a member's account are paid out first. This method can be used as an accounting system for inventory as well.

First mortgage lien

A lien is a legal right to control or enforce a charge against a specified property until some legal claim is paid or otherwise satisfied. A first mortgage lien is superior to any other security claim on the same property. The soundness of a first lien, for investment purposes, depends primarily on the value of the property being mortgaged. In rural electrification, CFC and REA cohold first mortgage liens, on a prorated basis, as security for a concurrent loan.

Front-end financing

Financing required by a rural electric system, generally a power supply system, before long-term financing becomes available. This financing may be required for such purposes as feasibility, engineering, or environmental studies; site acquisitions; or payments in connection with fuel supplies and equipment.

Generation and transmission cooperative (G&T)

A cooperative supplier of wholesale power to distribution systems or to other G&Ts.

Guarantee

A contract or agreement involving three parties. The first party, CFC for example, agrees to assure that a second party, a CFC member system, meets its obligations in accordance with the contract or agreement to benefit the third party, the creditor or investor. CFC provides its guarantee generally to member G&T systems in connection with the issuance of tax-exempt bonds to finance pollution control facilities and to support lease transactions. With CFC's guarantee, members can usually get a better rating on their bonds and thus obtain financing at a lower interest cost.

Indenture

A document describing the types of securities and security provisions that may be pledged as collateral to support collateral trust bonds. The indenture also outlines standards that must be met to maintain securities' eligibility as collateral. The indenture gives the trustee the power to sell the pledged securities in a default situation and, from the proceeds, recover the sum due on the bonds. In the case of CFC's collateral trust bonds, pledged securities are represented by the notes and mortgages received by CFC, solely or shared with REA, on the facilities of its borrowers. The indenture sets forth maturity date, interest rate, security and other terms. To raise funds to loan to its members, CFC pledges as collateral the notes and mortgages of its borrowing systems.

Insured loans

A loan made by the REA directly to the rural electric borrower. The loan is secured by a mortgage on the borrower's facilities. These loans are made from the Rural Electrification and Telephone Revolving Fund set up for REA's use. REA is authorized to use the security of the mortgage in the revolving fund as security for the sale of notes on certificates of beneficial ownership to the U. S. Treasury (this being the way REA reissues capital).

Intermediate-term debt

Loans for a period of more than one year and generally not more than five years.

Investment banker

The principal medium for bringing securities to the investing public. The functions of the investment banker include the purchase of whole issues of securities from issuers and their distribution to institutional and individual investors, commonly called primary distribution.

Key ratio trend analysis (KRTA)

CFC's measurement of key ratios of distribution system performance sent out annually to rural electric systems. These ratios provide seven years of financial and operating information for a system. For comparison purposes, the KRTA shows median figures in 40 key financial and operational areas. It relates specific co-ops to others in the United States by state, size, G&T and growth grouping. The KRTA is a popular management tool.

Last in, first out (LIFO)

A method of retiring (paying back) capital credits to consumers. In this system, the most recent capital credits are paid first.

Lease

A sale of usage rights in an asset by one organization to another. In the lease transaction, one party—the lessor—grants to another party—the lessee—the rights of use, tenancy or occupancy of property. This property may be land, buildings, equipment or other chattel property. The lease agreement describes the rights of the owner (lessor) and the renter (lessee) and recites the terms of periodic payment and the tenure of the lease. The property leased reverts to the owner at the expiration of the lease agreement.

Letter of credit

An instrument issued by CFC to provide proof of financial backing for systems involved in projects requiring such backing. CFC provides letters of credit to systems involved in land reclamation or right-of-way projects. The letter of credit gives the system the strength of the financial backing of CFC, which issues the letter of credit in the system's behalf.

Leverage ratio

A measurement of an organization's outstanding debt relative to its equity (also referred to as a debt-to-equity ratio). CFC uses its member-invested equity as a lever by which it raises capital in the credit markets. Under terms of its indenture. CFC can raise $10 in debt capital for every $1 in equity, a 10 to 1 leveraging ratio. Subsequent amendments to the indenture will ultimately increase this ratio to 20 to 1. CFC's conservative debt-to-equity level contributes to its favorable credit ratings.

Leveraged leasing

A taxed-advantaged transaction in which *1)* the ownership of an asset tranfers from an organization with low taxes to one with a high tax burden, *2)* the asset's new owner then leases the facility back to the original owner and *3)* the new owner uses limited amounts of its own capital as equity in the transaction and borrows a large portion of the sale price. The original owner uses the facility at a lower cost than if it was owned outright. The new owner uses the various tax credits to achieve a lower tax bill.

Lien

The right of an organization or individual to control or to enforce a claim against property that it has an interest in, usually through a loan.

Lien accommodation

As used in the rural electrification program, a lien accommodation involves an agreement on the part of REA, which holds a first mortgage on all properties of rural electric system borrowers, to provide a first mortgage lien on the borrower's facilities equal to REA's lien on a prorated basis to another lender, such as CFC.

Line of credit

In a line-of-credit agreement between a bank (CFC) and a customer (a member), CFC agrees to lend the member funds up to a previously agreed on maximum amount. CFC has the option to withdraw from the agreement if the financial status of the borrower changes or if the borrower fails to use the line of credit for its intended use as per the agreement. The member may borrow as much of the line as is required and pays interest on the borrowed portion only and only for the period during which this portion is outstanding. A line of credit is widely used by large organizations to fund future commitments and purchase inventory. CFC requires periodic financial reports from the borrower systems so as to be constantly informed on their credit status.

Loan

A business transaction between two "legal entities," where one, known as the lender, agrees to "rent" funds to the second, known as the borrower. The funds are almost always rented for a fee, called interest. Banks are the principal lenders of funds for commercial purposes. Loans may be demand or time loans, depending on the agreement as to maturity. They may also be secured or unsecured within the above categories. When secured, the loans are supported by assets acceptable to the lender, and these assets may be used by the lender to recover payment in case of default by the borrower. Loans may also be short term (less than 1 year), intermediate term (1 to 5 years) or long term (more than 5 years).

Loan programs

CFC provides a full range of loan programs to meet rural electrics' varying capital needs:

Most CFC **long-term loans** are made to distribution members concurrently with long-term loans from REA. These loans are generally made for a term of 35 years and are used by members for constructing electric plants and related facilities and for acquiring equipment, materials and fuel supplies. Concurrent long-term loans are secured by a CFC-REA common mortgage. CFC also provides 100 percent long-term loans to members for electric plants and related facilities. Such loans are usually made when REA funds are not available due to limitations placed on the agency's lending authority.

Intermediate-term loans generally are secured loans, made for up to 5-year periods, and are used primarily by G&T systems for front-end financing of power supply projects and as interim financing for lease transactions until long-term financing can be obtained.

Members use CFC's **short-term lines of credit** for a wide variety of general corporate purposes. Line-of-credit commitments are made for periods up to 12 months. After a member establishes a line of credit, funds may be drawn and repaid as often as necessary, as long as the amount outstanding at any one time does not exceed the total amount of the approved line.

National Cooperative Services Corporation (NCSC)

A financial service organization established in 1981 by CFC to develop more efficient ways to finance large power supply projects and arrange leases. NCSC employs the services of leading investment bankers and, on request, participates in and structures lease transactions and other specialized financing projects for rural electric systems.

Paying agent

A commercial bank or CFC can be charged with the responsibility of making any and all payments of interest and/or principal or other bills in behalf of a customer, or in CFC's case, a member.

Percentage method (capital credits retirement)

The retirement (payback) of capital credits to consumers based on a percentage return of all capital credits in that consumer's account. This repayment form ignores the time that money has been in the account and does not return patronage based on FIFO.

Plant revenue ratio (PRR)

A measurement of the relative productivity of a plant of a borrower. PRR is used to determine, for borrowers required to obtain supplemental financing, the percentage (10, 20 or 30) of each long-term, concurrent loan that must come from a supplemental source, such as CFC.

Private placement, or private offering (in the private-capital market)

The sale of securities through negotiations with a buyer who purchases the securities for investment purposes without registration of the securities with the SEC.

Revolving fund

A fund that uses repayments on loans plus existing assets to make new loans, thus establishing its revolving nature. REA's revolving fund, established in the early 1970s, was created through an $8 billion appropriation by Congress of loans and other assets. The fund used these loans and assets to make future REA loans.

Rural Electrification Administration (REA)

The federal agency in the U.S. Department of Agriculture that deals generally with electric membership corporations. Among other functions, the REA approves and guarantees loans to electric cooperatives for generation, transmission and distribution facilities. Congress passed the Rural Electrification Act establishing REA in 1936.

REA guarantee

Credit support from the REA that provides a guarantee to the FFB stating that the REA will pay the FFB, from the revolving fund, for loans that are in default.

Rural Telephone Finance Cooperative (RTFC)

A not-for-profit financial services corporation created by CFC in 1987 for the purpose of providing private-market capital to rural telephone systems and their affiliates.

Secured loan

A loan that is secured by a mortgage of property, a pledge of marketable securities, or other valuable property or rights that can be sold to satisfy the debt on failure of the borrower to pay.

Securities and Exchange Commission (SEC)

The governmental body that administers federal security laws through a five-man commission appointed by the president. Established under the terms of the Securities Exchange Act of 1934 and the Securities Act of 1933 to protect the interests of the investing public. It supervises the registration of security issues and other transactions under terms of the Securities Act.

Short-term debt

Borrowings that are payable within one year. CFC's primary sources of short-term debt include sale of commercial paper through CFC's commercial paper dealer and to members through CFC's Money Desk, direct short-term borrowings from commercial banks and master notes issued to bank trust departments.

Swap

A financial transaction whereby domestic and foreign organizations trade responsibilities to make interest payments on each other's debt. A swap takes advantage of the differences between long-term and short-term interest rates in each country, benefiting both.

Tax-exempt bonds

A term applied to revenue or general obligation bonds of state and local governments or agencies. The interest on municipal securities is exempt from federal income taxes and some state taxes.

Times interest earned ratio (TIER)

A financial measurement relating to the borrower's ability on an annual basis to earn margins sufficient to cover the interest charges on its total outstanding indebtedness (long and short term). CFC's criterion for pledgeability of a distribution system's mortgage as security for a bond issue requires a TIER of 1.50 or more, which means the borrower must have annual margins equal to 150 percent or one and one-half times the annual interest cost on total indebtedness. CFC requires systems to maintain an average TIER of 1.50 of the two highest of the last three years.

Warehousing loans

The act of funding long-term loans with short-term debt until the total amount is of sufficient size to justify a bond issue to be secured by long-term loans.

INDEX

D

E

J

K

L

U

V

W

Y